The 1st and 2nd Stuart Horse Artillery

The Virginia Regimental Histories Series

1st Edition

This series is dedicated to the State of Virginia and all of her people who lived during the War Between the States. It is the purpose of this series to preserve, as a part of our heritage, the deeds and sacrifices of these Virginians. Your support of this project is greatly appreciated.

Number 527 of 1,000

Robert H. Moore, II

Manufactured in the United States by
H. E. Howard, Inc., Lynchburg, Virginia

Printed by H. E. Howard, Inc.

ISBN-1-56190-108-3

The 1st and 2nd Stuart Horse Artillery

The Virginia Regimental Histories Series

1st Edition

Robert H. Moore, II

The 1st and 2nd Stuart Horse Artillery

Robert H. Moore, II

DEDICATION

Dedicated to the men of the Horse Artillery of the Army of Northern Virginia;
the "Long Arm" of Lee's Cavalry.

And to my wife,
Danette,
who stood by me with patience
while I struggled forward with the guns and caissons just one more time.

Acknowledgments

For more than ten years, the Virginia Regimental Histories Series has been a part of my life. For others, the series has been a companion far longer. I think that most will look back at their works for the series with a sense of satisfaction. Through it, I personally have become acquainted with many who shared an interest in the units and more especially the artillery from Virginia. Likewise, though silent observers from some unseen plain, many of the men who actually served in the various batteries rekindled the spirit of their units through letters, diaries, photos, brief reminiscences and various other means. Though I did not personally ride the caissons or serve the guns with Virginia's gallant sons of Saint Barbara, I feel honored that I could pick up and complete a unit history which many veterans themselves had only hoped would be completed. It is the deeds of these men and the gallantry of these artillery batteries that needs to praised first and foremost. I can only hope that I have done them some justice.

Mr. Harold E. Howard has been the backbone of this series and deserves tremendous applause for sticking out the effort that expanded even beyond his imagination. I think all that have written for the series can agree with Harold's idea that the project was all a "labor of love." Likewise, while there "would be no book before its time," Harold also had to contend with delays and promises that the manuscripts were on the way--my own manuscripts being no exception.

My dear wife, Danette, most certainly must have come to the same conclusion that Harold did when I would proclaim as I neared the completion of each book that "I'm almost done with it." She has been a champion throughout and has provided support and understanding for my time on the computer and "in the field" digging up the letters and such, copying rosters, and even trying to find where the batteries were actually located during the various battles. She too has braved the microfilm readers for the same and is to receive equal satisfaction in the completion of this and previous works. I thank her and am grateful for her love and patience.

Mr. Robert K. Krick of Fredericksburg, as ever, deserves thanks for allowing me to use microfilm and his personal files on Virginia Units. R. E. L. Krick of Richmond also contributed to this work from his files concerning newspaper articles on the men and units. Mrs. Susan M. Collins and the late John Ingalls and the staff of the Jefferson County Museum at Charles Town, West Virginia , deserve a special thanks for their fine support in allowing the use of the invaluable Chew manuscript collection and the continuing support in the final days of the preparation of this manuscript. My life-long friend Eric D. Feid, now of Irving, Texas, is to be thanked for assisting with miscellaneous materials pertaining to John Pelham, photos of our numerous adventures of years past to the sites and battlefields, and memories of those same battlefields where many a caisson rambled over. A very hearty thanks to Peggy Vogtsberger and the John Pelham Historical Association for providing assistance and newsletters for several years prior to this manuscript actually getting off the ground. Thanks are also extended to Mr. William H. Strong, Jr. for the photo and information concerning his 1st Stuart Horse Artillery guidon. Thanks also to Robert G. Phlegar of Salem, Virginia, for contributing a photo and information concerning his great-grandfather.

Other organizations and their respective staff of archives and libraries that deserve thanks include the Virginia Military Institute, Lexington, Virginia; United States Military Academy, West Point, New York; the United States Army Military History Institute, Carlisle, Pennsylvania; The Library of Congress, Washington, D.C.; Houghton Library, Harvard University, Cambridge, Massachusetts; The Museum of the Confederacy, Library of Virginia and Virginia Historical Society Library, Richmond, Virginia; Duke University, Durham, North Carolina; University of Kentucky, Lexington, Kentucky; and The University of Virginia, Charlottesville, Virginia.

And of course, to my good friend, Mr. Jeff Weaver, for rallying in assorted ways of assistance in the last days of the preparation of this volume.

Finally thanks are given to God for having seen me through the rocky beginnings in the last phases of this project on the horse artillery batteries and for continuing to see me through the projects to come.

Robert H. Moore, II
Kingsland, Georgia
January, 1998

PREFACE

Introduction to the Organization of the Horse Artillery Batteries in the Army of Northern Virginia

On November 13, 1861, the first horse artillery company in Virginia and the Confederate service was organized under the influence of Colonel Turner Ashby at Flowing Spring, Jefferson County, Virginia. Serving initially as a company of the famed 7th Virginia Cavalry, Ashby's Battery sounded the revival of the use of horse artillery as an effective element in warfare.

Typically, horse artillery was thought to be ineffective because it caused a great demand upon the horses and was often considered a "waste of good horse flesh." The idea, however, was not reserved to Colonel Ashby, but was appealing to other cavaliers as well. Within days of the organization of the Ashby Battery, General J. E. B. Stuart formed his own horse battery from the remnants of the Newtown Artillery.

The following summer Ashby's and Stuart's idea would produce spectacular names in military history such as Chew, Pelham, Breathed, and Beckham. By the close of the summer of 1862 Stuart's Horse Battery, by then known as Pelham's Horse Artillery, would form the base for an artillery battalion within the Cavalry Corps of the Army of Northern Virginia.

By September 2, Stuart's Horse Artillery was composed of James Breathed's 1st and Mathias W. Henry's 2nd Stuart Horse Artillery Batteries (both organized with commissions from Virginia) and James F. Hart's Washington (South Carolina) Battery. The battalion would continue to grow in the following months with the addition of Marcellus N. Moorman's "Beauregard" and William H. Griffin's 2nd Baltimore (Maryland) batteries. Though not an active participant with the battalion (acting mostly in the Shenandoah Valley), Chew's Battery would also be listed in its organization.

Upon the untimely death of Major Pelham in March 1863, the lead of the battalion was turned over to Major Robert F. Beckham, under whose reign the battalion continued to flourish in popularity and grow. During the summer of 1863, Thomas E. Jackson's Kanawha Battery would be temporarily added to the force for the Gettysburg Campaign and would continue to be intermittently listed with the several horse artillery battalions in the Army of Northern Virginia.

The additions of Warren Lurty's Battery in late 1863 and Edward Graham's Petersburg and McClanahan's Batteries in 1864 would bring the total number of batteries to serve in the Stuart Horse Artillery to ten. Before the end of the war James Dearing would be raised to the head the battalion, but never assumed command. Ultimately, that noteworthy post was given to Roger P. Chew.

Unlike the battalions of the infantry corps in the Army of Northern Virginia, the batteries of Stuart's Horse Artillery Battalion were not "fixed" to conglomerate service. Typically the different batteries would act independently of each other as they were assigned on different instances to the brigades of Stuart's Cavalry Corps. It was rare for all of the batteries of the battalion to serve together as a whole, such as at Brandy Station.

In 1864 the Stuart Horse Artillery Battalion was thought to be ineffectively integrated into the overall organization of the cavalry. By dividing the horse batteries into five battalions and assigning them to respective cavalry divisions, the paper and field organization of the multiple companies became easier to maintain.

The batteries that served alongside the cavalry of the Army of Northern Virginia proved the most active in the theater and perhaps the war. From Kernstown to Appomattox the batteries fulfilled the promise that the "Gallant" Pelham had once made that the service of the horse artillery "will be active and energetic, the men... will never be quartered in one place long at a time... [and] the career... is destined to be a brilliant one; and whenever or wherever there is likely to be a fight, they are bound to be in the front..."

Technical Background

Since artillery equipment used during the civil war is not generally well understood, the following tables give the basic equipment list and specifications for cannon used by Confederate artillerists during the Civil War. Descriptions of what each item was used for may be found in several books on weapons. A basic course in their function can be found in *Cannons, An Introduction to Civil War Artillery* by Dean S. Thomas. This volume is inexpensive, gets to the point quickly, and has illustrations of the equipment described.

The ideally outfitted light artillery battery had the following equipment for each gun:

sponges & rammers	2	sponge covers	2
worm & staff	½	handspikes	2
sponge bucket	1	prolonge	1
vent cover	1	tar bucket	1
water buckets	2	gunner's haversacks	2
tube pouches	2	vent pouch	1
gunner's pincers	1	tow hook	1
thumbstalls	2	priming wire	1
lanyards	2	gunner's gimlet	1
fuse cutter	1	tarpaulin	1

These items were scare in the Confederacy. Any battery with a full stock could consider itself lucky.

Almost every battery found itself equipped with a variety of armaments during the course of its history. Some of the terms may be unfamiliar to modern readers, therefore the following chart on the types of artillery is provided for reference:

Piece	Bore (in.)	Range (yds)	Weight (lbs.)
3-inch ordnance rifle	3.00	1830	967
6-pounder (bronze field gun)	3.67	1523	884
10-pounder Parrott	3.00	1830	899
12-pounder Napoleon	4.62	1680	1227
12-pounder mtn. howitzer	4.62	900	220
20-pounder Parrott	3.67	2100	1,750
24-pounder (bronze field gun)	5.82	1322	1318
8-inch sea coast mortar	8.00	1200	930
8-inch columbiad	8.00	1800	9,210

10-inch columbiad	10.00	1800	15,400

Each gun was typically pulled by a team of four or six horses and served by a crew of seven to ten men, distributed as follows:

In charge	sergeant
Gunner	corporal
Ammunition Chest	1-3 privates
Ramrod	1 private
Loader	1 private
Aimer	1 private
Firer	1 private

One or more teamsters would be present with the horses, or in the rear with them. On occasion, especially as attrition took men from various units, guns were served with smaller crews, with some loss in speed. David Cardwell, a member of the 2nd Stuart Horse Artillery, described some basic elements of a horse battery:

> A horse battery was generally equipped with four twelve--pound guns, sometimes Napoleons, sometimes three-inch rifled ordnance. Each gun was pulled by six horses and had a driver on each left--hand horse. There was a caisson to each gun, also pulled by six horses. The complement of men to each gun, according to the United States regulations, was eleven, and the "horse battery" differed from the Foot or mounted, having two extra men to hold the cannoneers' horses when they were dismounted to go into action. A horse battery maneuvered in action as frequently as practicable to save the men and horses when the enemy got their range too accurately.

A Brief Note to the Reader

This volume is the third of three as a part of the Virginia Regimental Histories Series to describe the roll of all of Virginia's horse artillery batteries and the batteries that served as a beginning for the same organizations, excepting McClanahan's Battery already completed for this series by Mr. Robert J. Driver. The history of the 1st and 2nd Stuart Artillery batteries is impossible to separate until the spring of 1864; therefore, the histories of the two have been combined in this volume into one history. Readers should take note, however, that in describing events starting in the spring of 1864, I have created separate chapters for the individual histories of the two batteries as they began to operate in two separate theaters of the war in Virginia. Chapters 6 through 9 cover that phase.

Evolution of the Stuart Horse Artillery Battalion: 1861 - 1862

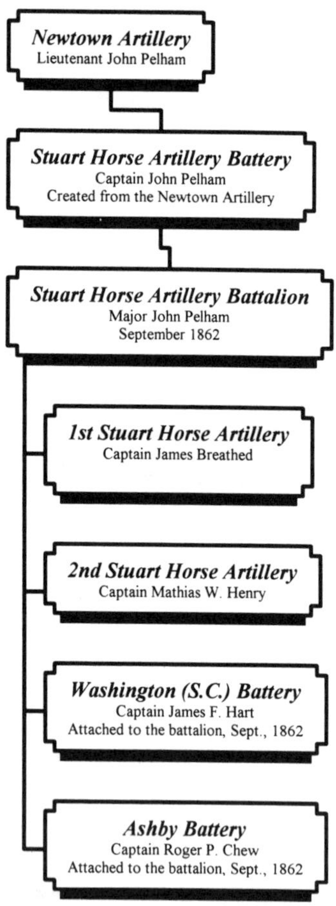

Stuart Horse Artillery Battalion: Summer 1863

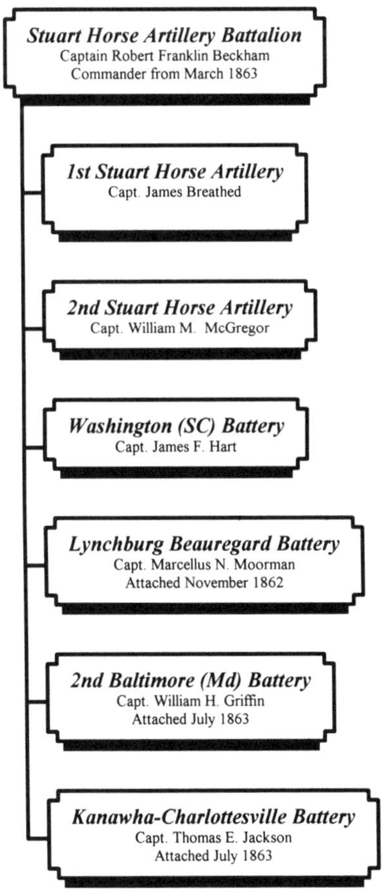

Stuart Horse Artillery Battalion: October 1864

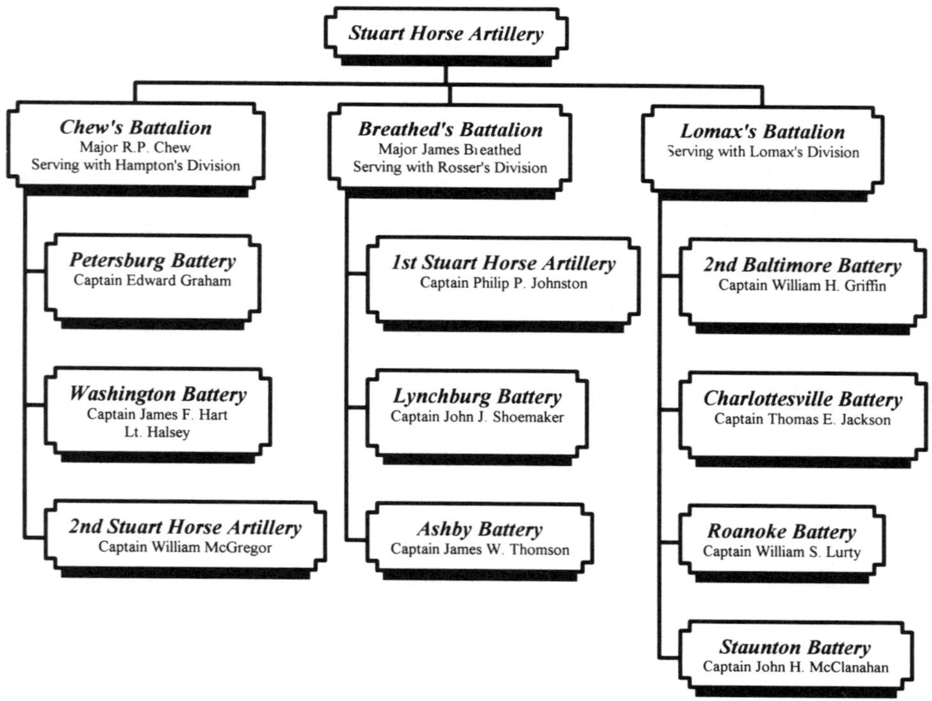

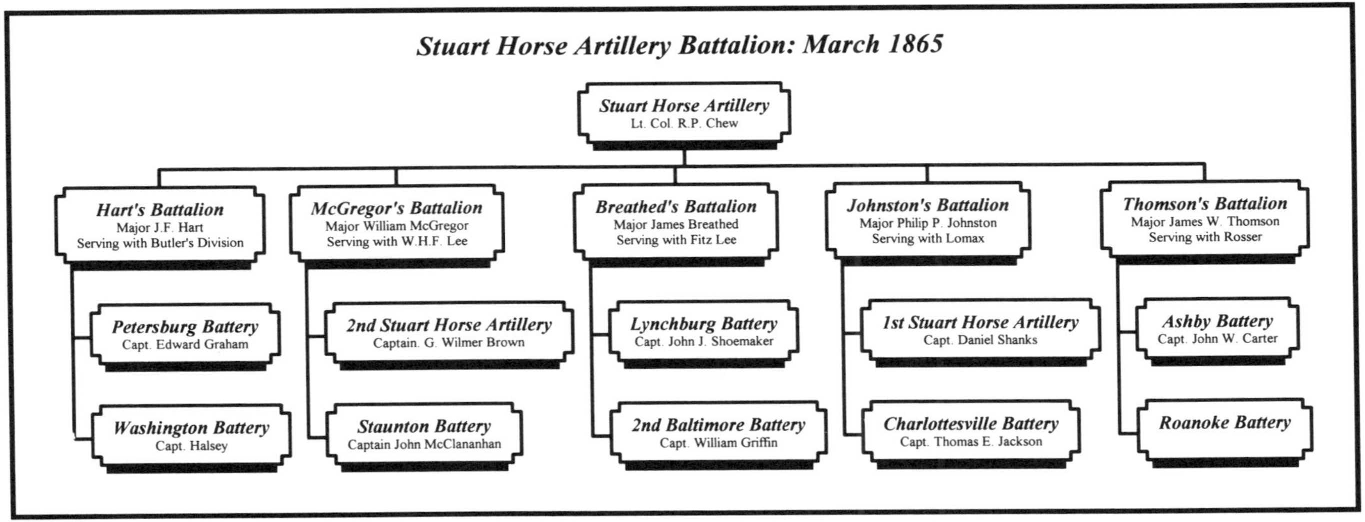
Stuart Horse Artillery Battalion: March 1865
Stuart Horse Artillery
Lt. Col. R.P. Chew
Hart's Battalion
Major J.F. Hart
Serving with Butler's Division
Petersburg Battery
Capt. Edward Graham
Washington Battery
Capt. Halsey
McGregor's Battalion
Major William McGregor
Serving with W.H.F. Lee
2nd Stuart Horse Artillery
Captain. G. Wilmer Brown
Staunton Battery
Captain John McClananhan
Breathed's Battalion
Major James Breathed
Serving with Fitz Lee
Lynchburg Battery
Capt. John J. Shoemaker
2nd Baltimore Battery
Capt. William Griffin
Johnston's Battalion
Major Philip P. Johnston
Serving with Lomax
1st Stuart Horse Artillery
Capt. Daniel Shanks
Charlottesville Battery
Capt. Thomas E. Jackson
Thomson's Battalion
Major James W. Thomson
Serving with Rosser
Ashby Battery
Capt. John W. Carter
Roanoke Battery

Chapter 1

The Stuart Horse Artillery Battery

"I have got the finest equipment and the finest guns in the service and I want good men to man them."

1861

In early August 1861, General Joseph E. Johnston recommended the organization of a brigade of cavalry and suggested that Colonel James Ewell Brown Stuart be promoted to lead it. During the process of organization, Stuart made it clear that he desired a section of horse artillery for the support of the new brigade. In his proposal, Stuart recommended that two men be transferred from each company. The two guns for the new section would be provided by Governor John Letcher of Virginia. However simple it may have seemed, the proposal met opposition from Secretary of War Judah Benjamin and President Jefferson Davis. Furthermore, Stuart's recommendation for John Esteen Cooke as commander of the section was met with a lack of enthusiasm. Cooke had been a former sergeant and chief-of-piece for the Richmond Howitzers and seemed to Stuart the most likely candidate for the position. Cooke was not approved by the government for the post.

Another candidate was then raised in James W. Breathed. Having served in the 1st Virginia Cavalry as a private, Breathed had impressed Stuart much. After leaving Fort Riley, Kansas, in April, Stuart first met Breathed on a train ride from Memphis in early May. A physician and native Marylander, Breathed was little experienced as an officer. His early life was spent at a military school in Urbanna, which had been run by future Confederate general John Robert Jones. From there, the young Breathed attended St. James College and the University of Maryland Medical College. After receiving his degree in medicine in 1858, the Marylander headed west and opened a practice in St. Joseph, Missouri.

Nevertheless, Stuart had developed enough admiration and trust in Breathed to take the initial steps in finalizing a commission. By November 18, the young gentleman from western Maryland held a lieutenant's commission in the 1st Virginia Cavalry. Two days later Stuart had his new officer transferred to his artillery, which in fact was still only on paper. Though taking immediate strides to get his artillery officer appointed, Stuart still received resistance from Richmond.

Finally, a recommendation was forwarded to Stuart proposing a more widely accepted candidate, Lieutenant John Pelham. Pelham had already distinguished himself as a worthy commander. The Alabamian had slowly been working his way into command. A fellow West Point classmate, Edward Porter Alexander, remembered Pelham as "a very young looking, handsome, & attractive fellow, slender, blue eyes, light hair, smooth, red & white complexion, & with such a modest and refined expression that his classmates & friends spoke of him as "Sallie" and there never was a Sallie whom a man could love more!" Having left West Point prior to the graduation of 1861, the young Calhoun County native found his way to a commission with the assistance of some well placed letters written by his father, Dr. Atkinson Pelham. Having actually received his commission as lieutenant of artillery in the Provisional Army of the Confederate States on March 16, Pelham first saw service with the Wise Artillery of Martinsburg. While the artillerists were being reorganized in June, twenty-two year old 1st Lieutenant Pelham was detailed as drill master. When he arrived, the fuzz-faced boy was probably greeted with little enthusiasm by the 44-year old Captain Ephraim Alburtis. Tactfully offering his services, Pelham soon had the company for his own to drill as he deemed necessary. In the days that followed, the inexperienced company gained heavy training in the fields between Winchester and Martinsburg. In a little over a month, Pelham would distinguish himself in battle for the first time at Manassas. Pelham made the rounds after the battle, serving briefly with Frank Schaeffer's Beauregard Rifles and finally ending up as the commanding officer, as a lieutenant, of the Newtown Artillery. Throughout, the blond-haired artillerists never lost his zest for action and was thoroughly active in social circles. Wrote one member of Stuart's staff, he was "as grand a flirt as ever lived."

Stuart, having witnessed Pelham in action at Manassas, supposedly made the comment that "next to Rosser probably no graduate [of West Point] would be better." By November 29 it was finalized by General Joseph E. Johnston in Special Orders No. 557. Pelham was transferred "for service as horse artillery." Pelham, being at the time still the commander of the nearly defunct Newtown Artillery, brought with him whatever remained of the old battery, including a few men that were left.

Since the early rosters of the Stuart Horse Artillery are seemingly far from complete, the exact number of men that transferred from the Newtown Artillery cannot be certain. The only two men, apart from Pelham, that appear active on the Stuart Horse rolls by 1864 were Dennis Conroy and Isaac L. Holmes. Conroy and Holmes were devoted artillerists who remained with the battery to the last days. Conroy even made the extra effort to join Joseph E. Johnston's army in 1865, after the surrender of Lee. It was with Johnston at Greensboro, when Conroy finally surrendered on April 26.

Though the initial steps for organization of a horse battery had taken place there was much work ahead. The former battery from Frederick

County, Virginia had supplied a few men and a fair number had transferred from the cavalry; however, much recruiting remained necessary.

In December 1861, a group of recruits from Alabama was gathered just prior to Pelham's actual transfer to Stuart. It is most likely that the bulk of the forty Alabamians were gathered when Lieutenant George Wilmer Brown made a trip to the Pelham home at Alexandria, Alabama.

Lieutenant Brown himself was no insignificant character in the organization of the Stuart Horse Artillery. The son of Reverend Richard Templeton Brown, Wilmer came from old Virginia and Maryland blood, being a descendant of one of the original Jamestown colonists. His grandfather was William Holland Wilmer, the former president of the College of William and Mary. Wilmer entered the Virginia Military Institute in 1859. At the opening of the war he was sent with the corps to serve as a drill master. He was assigned to a company of the 15th Virginia Infantry. Joining Pelham's battery as a private, Brown soon found himself with a commission as a lieutenant of artillery.

On December 19 a recruitment poster appeared in the *Jacksonville Republican:*

> Mounted Flying Artillery
>
> CAPT. JOHN PELHAM, of Calhoun County, who is now in the Confederate army at Manassas, temporarily in command of the Jeff Davis Artillery, is authorized to raise two hundred volunteers for Mounted Artillerists, to serve during the war. The Confederate States will furnish 240 horses, and eight splendid brass cannon, wagons, caissons and equipments complete for men and horses.
>
> This flying artillery, when raised and organized, is to have one Captain, 3 first and a second Lieutenants, and non-commissioned Officers in proportion, and all will be attached to General J.E.B. Stewart's Cavalry.
>
> The artillery of 8 splendid brass guns - 200 men rank and file, and 250 horses, has the novelty of being the only one in our service - in fact the only one in America. All the men or Cannoneers are mounted on horses furnished by the government.
>
> This doubtless will be the most desirable, pleasant, and efficient service in our entire Army. This service will be active and energetic; then men will never be detailed to

> guard forts, fortifications, baggage or prisoners; neither will they be quartered in one place long at a time. The career of this company is destined to be a brilliant one; and whenever or where ever there is likely to be a fight, they are bound to be in the front - sustained by the gallant General Stewart....
>
> If there be any young men who desire to enlist for the defence of their native sunny South - who desire eminent distinction, Glory or renown, here is no doubt the finest opening that has yet been offered....

Lieutenant Brown made his enlistment headquarters at Pelham's father's home in Alexandria and at Jacksonville where he could be "consulted in reference to this important service." In the closing of the article, an extract from one of Pelham's letters was entered:

> Now is the time to serve your country - enlist the interest of The Ladies - tell them I want to do something to render myself worthy of them, and they must aid me in furnishing men. I have got the finest equipment and the finest guns in the service and I want good men to man them.

In his biography of Pelham, Milham wrote that the Alabama contingent was a "colorful one. Some twelve or fourteen of them were from Talladega, enlisted with the help of one of Pelham's boyhood friends, William McGregor, a Talladega lawyer; while the remainder came from Mobile, with a sprinkling from New Orleans. French and Italian they were by birth or immediate heritage, chiefly French." These men were later referred to Pelham often as the "Napoleon detachment." When the men arrived in Virginia to join the battery is unclear, but they apparently joined the company at Culpeper Court House.

McGregor would continue to play a significant role throughout the history of the Stuart Horse Artillery. A graduate of Auburn University, he had first served with Company E of the 10th Alabama Infantry. However, with the influence of his old friend Pelham, the Talladega native soon found his way into the ranks as a lieutenant in the Stuart Horse Artillery.

Armed initially with the 6-pounders used at the battle of 1st Manassas, the battery had much growing to do. By December 20 the battery still was far below numbers at the time of the action at Dranesville. One day following the engagement the company listed 94 horses on the rolls. The men certainly would find their compliment sufficient to sustain serious service during the following year. In the meantime, the men prepared for their first winter in the field and began to build winter quarters at "Camp Qui Vive" near Culpeper.

Chapter 2

Spring & Summer 1862

"The career of this company is destined to be a brilliant one...."

Just after the opening of the year, the young battery reflected growth when it reported 100 horses present in a January 7 requisition for feed. However, poor equipage was also reflected in the fact that requisitions were made for lengths of rope in lieu of proper harness supplies. Despite the inadequate supplies, drilling continued daily. However, unlike their other Southern brethren in the service of Saint Barbara, Pelham's men had an escape from the tedium of basic artillery drill as the refined art of horsemanship also had to be mastered.

Things began to look up in supply matters when Governor Letcher, through cooperation with Colonel William Nelson Pendleton, saw to it that the battery was augmented by twelve pounder howitzers. One of the guns nearly never made it to the company. While the piece was en route, Captain S. P. Hamilton of Georgia "procured" it. However, on March 5, Stuart made certain the gun continued on its journey to his company and wrote Pendleton of his insistence that "that identical gun and no other must be restored to the *'Stuart Horse Artillery'*."

As spring drew near, Pelham began to seek new recruits. To encourage the enrolling process, young Lieutenant Pelham made a quick trip to Richmond and obtained $600 in bounty money. Charles Ford made a great recruiting effort on behalf of the battery with the monetary incentive and by March 23, the roster had reached sufficient numbers of men in order to proceed with official organization. In that organization, Pelham was selected as captain of the battery; James Breathed as 1st lieutenant; William McGregor as 2nd lieutenant; William Elston, 3rd lieutenant; and James T. Shepherd as 4th lieutenant. Though elected to their posts in the battery, the commissions were not official until granted by Governor Letcher beginning on May 1 with the confirmation of Captain John Pelham as an officer in the Provisional Army of Virginia. Thus, the battery would be recognized as a Virginia battery, despite its numerous volunteers from other states.

Non-commissioned officers included W. S. Dabney, 1st sergeant; Ashton Chichester, 2nd sergeant; Joseph Holmes, 3rd sergeant; J. Cook, 4th sergeant; Daniel Shanks, 5th sergeant; Charles E. Ford, 6th sergeant; and G. Wilmer Brown, 7th sergeant.

Charles Ford's sister had, the previous fall, brought Stuart some information of value. In recognition of her success as a spy, Stuart issued the following document lauding her patriotism:

> To all whom it may concern
>
> Know ye
>
> That reposing special confidence in the patriotism, fidelity and ability of Antonia J. Ford, I, James E. B. Stuart, by virtue of the power vested in me as Brigadier General in the Provisional Army of the Confederate States of America, do hereby appoint and commission her my Honorary aid-de-camp, to rank as such from this date.
>
> She will be obeyed, respected and admired by all true lovers of a noble nature.
>
> Given under my hand and seal at the Head Quarters of The Cavalry Brigade at Camp Beverly, the 7th October, A.D. 1861 and the first year of our independence.
>
> J. E. B. Stuart

In 1863, Antonia's practice as a spy would cause her arrest and a great deal of concern to Stuart.

On April 9, the Stuart Horse Artillery was well on the way to the Peninsula with the rest of Stuart's Brigade. Though small at first, this little grouping of cavalry included the 1st and 4th Virginia Cavalry, the Jeff Davis Legion, and the Stuart Horse Artillery. Along the route to join General John Bankhead Magruder at Yorktown, the young artillery company was still in desperate need of men and supplies.

During the march from Culpeper to the Peninsula, the company gained a large number of recruits while passing near Richmond. There, meeting with a detachment of raw recruit militia from Floyd County, Pelham's ranks swelled to even greater numbers. Supply-wise, the company faired well also. On or about April 16 the company acquired 50 new horses, all of which, however, needed to be trained for battery use. Likewise, and due to the conscription law, nearly another half-dozen recruits joined Pelham's company. In the reports of strengths for organizations during the Peninsula Campaign, the Stuart Horse Artillery was listed with 141 men. The company also had 130 horses, two 6-pounder Howitzers, two 12-pounder Howitzers, one Napoleon, and one 3-inch Blakely rifle.

Map, Fort Magruder

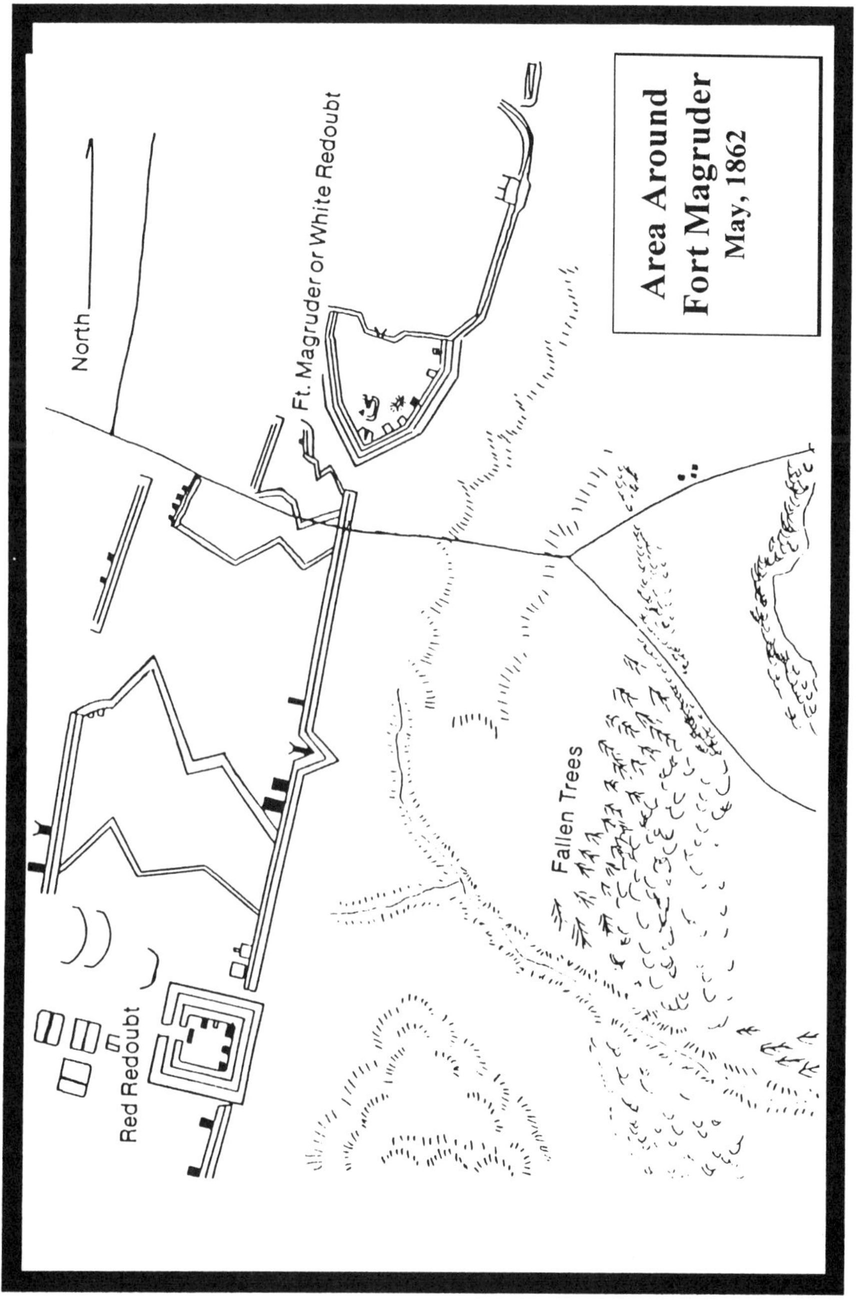

When speaking of the mix of men in the Stuart Horse Artillery, George Shreve wrote:

> Our Battery was very cosmopolitan in its make up. We enrolled men from Maryland, Virginia, South Carolina, Georgia, Alabama, Mississippi, Louisiana, and Tennessee. A large majority were volunteers. We had, however, quite a number (perhaps ten or fifteen) of conscripts from South west Virginia. These made very good soldiers, but lacked the enthusiasm that characterized the others. In fact, some of them were bitterly opposed to war, and a few were excused and allowed to return home on account of their religious scruples. Some men who were very timid at first, afterward, made good soldiers. I have in mind, a boy about nineteen who was in the first battle, in which we took part, in serving the gun, ran away some fifty feet or more, from the gun, and could hardly be persuaded to return, being so frightened. Sergeant Chichester went after him, and placing himself behind him, with both hands on his shoulders, forced him forward to his post, and threatened to shoot him if he deserted it again. He made a good soldier, and never showed cowardice after that.

One of the interesting figures to come to the Stuart Horse Artillery during the spring of 1862 was Private James C. Murdock. Having recently transferred from the 1st Virginia Cavalry, Murdock, a native of Frederick County, Maryland, was a well educated man, having attended the University of Edinburg in Scotland prior to the war. Though there is little of his service with the Stuart Horse, it was Murdock's service after the battery that was interesting and unusual. After only serving with the company for less than a year, Murdock received a commission as 2nd lieutenant in the Confederate States Marine Corps on April 8, 1864. First serving on the C.S.S. *Richmond*, Murdock was quickly transferred in May to the works at Fort Fisher, near Wilmington, North Carolina and was captured in the assault upon that place on January 15, 1865. After being exchanged in March, Murdock left little time for rest and was quickly assigned to the James River Squadron. In the closing months of the war, the 25 year old lieutenant distinguished himself in assisting in the capture of the steamers *St. Mary* and *J. B. Spafford* in Chesapeake Bay on March 31. Murdock would finally close out his service when he received a parole at Society Hill, South Carolina on May 27, 1865.

The ordnance for the Stuart Horse Artillery was typical of the Confederate artillery service with one exception. The Blakely gun was a rare item that had just recently come into service. Developed in the late 1850's by former Royal Artillery Captain Alexander Theophilus Blakely, the cannons were constructed of steel or cast iron and banded with steel. Firing a 1.5 pound powder charge "bolt," these guns had an effective range of up to 1750 yards. By the spring of 1862, the most recently received pieces had been purchased by foreign agents and brought into the Confederacy by a fleet of blockade runners operated by the Bureau of Ordnance. Manufactured in England, the Blakely gun would become a familiar piece to the horse artillery for sometime. Though a fine piece of artillery and lightweight for horse artillery use, the Stuart Horse Artillery would find reoccurring problems with it throughout the year, most especially from the fact that the carriage was not properly constructed to handle the severe recoil of the gun. Later in the war, the guns were diminished in favoritism more when suitable ammunition became hard to obtain. As a consequence and by the following year, the guns were slowly phased out of the Stuart Horse Artillery Battalion.

Though he enjoyed the new growth, Pelham, if only briefly, was severely criticized for his methods of supply that April. The source of the criticism was the Quartermaster General, Colonel A. C. Myers, for actions taken en route to Yorktown. Myers protested to General Johnston that Pelham, "when not more than two miles from Richmond on his march to Yorktown, he without the color of propriety or necessity, seized two valuable mules and harness from the wagon of a citizen of Henrico County, which was met in charge of a servant." Colonel Myers elaborated on the incident:

> Captain Pelham was supplied fully with the means of transportation and was not compelled by any want of animals to commit this wrong. He gave the citizen (a Mr. Quarles) a certificate that he had received the mules; but made no mention of the gearing. I directed Major Barbour to settle with him in the value of the mules, but could not, in the absence of the necessary certificate, direct the gearing to be included. I regard this as a wanton act of oppression on the part of Captain Pelham, and as such report it to you.... The burdens of war are severe enough upon our people, without their being made more onerous by the arbitrary conduct of military officers....

In response to the charges, Stuart was more than happy to thunder back a sharp reply. In his heated explanation Stuart clarified the matter and stated that "while on the march, the roads through Culpeper were found very difficult and impassable by wagons." Further details are best described in Stuart's exact words:

> Discarding baggage I resorted to pack mules and accomplished this march to Richmond, through rain, snow & sleet without tents or other shelter except such as could be im-

> provised from nature. My Adjt General with great difficulty obtained one days supplies to be hauled from the city to my camp the first day--subsequently however transportation for commissary stores & ordnance was peremptorily refused the Asst Qr Mr & wagon master saying that his orders were positive not to allow any transportation to Gen Stuart's Camp. As my command was the only one near Richmond without wagons (which was distinctly set forth) this discrimination seemed unaccountable, but was persisted in, the Qr Mr saying we "had no business coming without wagons." My application was repeated and urgent but Maj Larkin Smith's orders were peremptory. The necessity for impressment of the necessary transportation was as clear and absolute as any military necessity could be; and I do not hesitate to resort to it, though when practicable as in the case of the Hon Mr Lyons I obtained the owners consent. The wagons were all returned except one which owing to the delay in obtaining ammunition and some necessary repairs to pistols was retained to overtake me on the march. It happened to be that of the Hon Lyons, which I consider most fortunate, as none of the others were as able to submit to the temporary inconvenience as himself.

Continuing to severely chastise the colonel, Stuart also brought up details about Pelham's role:

> Upon reaching Richmond his command was increased 100 men — without any increase of transportation. My experience certainly forbade the attempt to procure any from the QM Dept at Richmond. With his usual perseverance Capt P started with what he had but before proceeding far found it absolutely necessary to procure two mules & harness giving the owner assurance of restitution or ample compensation — receipting for the property.... If purchasing of Mr Quarles at full value his mules be a hardship on the people of Henrico they have much to learn. A people who have literally fattened on the war may well bear some of its burdens. Yet our artillery must stick in the mud while the streets of Richmond are thronged with spanking teams, enjoying immunity, when the very act of wanton oppression complained of was a measure directly tending to the defence & protection of the city & with it the inviolable chattels of Messrs Lyons & Quarles. Col. Myers sweeping declaration against Capt Pelham amounts to nothing, for you & I know what a high-toned, zealous, and efficient officer he is.

In the years following the war, Pelham was also criticized by John Cheves Haskell. Haskell wrote that though Pelham "early distinguished himself by fighting close and hard... he was not a good organizer and disciplinarian, and seldom had his command in good condition." It is interesting to note that the records of the 1st and 2nd Stuart Horse batteries are extremely poor up until after Pelham's death. The Combined Service Records do not accurately maintained until 1864. Other than "poor organization," there are at least two other explanations regarding the record keeping. First, the reports and records may have been lost in the destruction and evacuation of Richmond near the close of the war. The other area to consider is the nature of the service of the battery, and later, batteries. These units, unlike their "foot artillery" counterparts, were rarely in camp long enough to maintain proper administrative records. When they were in camp for any long amount of time, the need to resupply horses and equipment seemed to dominate the matters of business.

Even this was criticized when Haskell stated "If he had been confined for his supply of horses to those drawn through the quartermaster's department, he would have been scarcely ever able to transport half his guns." Haskell seems to show a bit of jealousy towards the horse artillery when he followed up with the statement; "But moving constantly with the cavalry and always at the front, his command picked up fresh horses and equipment, which never passed through the hands of the department." Still convinced he was a poor organizer, Haskell concluded that "Even with his extra supply, he was often unable to move all his guns when he wanted, but his rule was to be done and to wait for no stragglers...."

What Haskell seemingly doesn't take into account was the very nature of the service of horse artillery. As earlier mentioned, the horse artillery was often considered a "waste of good horse flesh," and it did wear down mounts at an alarming rate. Throughout Pelham's service, the 1st and 2nd Stuart Horse Artillery were rarely fielded with full strength in battle. However, with what little he did have, even Haskell was forced to admit that Pelham was the first who ever demonstrated that artillery could and should be fought on the musketry line of battle; and that the men of the batteries were "dashing and effective fighters."

By April 21, Stuart's Brigade was in camp at Bigler's Wharf near Yorktown. However, the brigade and its artillery did not sit idle for too long, moving on April 28 to Green's Wharf on the James River. On May 2, Pelham's Battery moved out of Yorktown and reported to Colonel Stephen D. Lee at Bigler's Wharf to help guard against possible landings of Federal troops from transports. While the bulk of the battery stayed at Bigler's, 2nd Lieutenant James Shepard was ordered to accompany Stuart's Brigade with a section of mountain howitzers. The section of guns spent the next few days posted at strategic points between Williamsburg and the Yorktown-

Warwick River Line. However, with General George McClellan's massive Federal army bearing down on Johnston, the Confederate general determined it best to evacuate the line by May 3.

On the same day that Johnston ordered the line on the Peninsula abandoned, Lieutenant Shepard, late in the afternoon near 4 p.m., took position near Colonel Williams C. Wickham's 4th Virginia Cavalry, on a small stream that empties near Grove Landing. Shepard remained there for several hours into the night. On the following day, while cavalry skirmishers protected the rear in the retreat, George Stoneman's cavalry made contact with the 4th Virginia and brought Shepard's howitzers into action. While in position just a mile or two below Fort Magruder on the Telegraph Road, the section fired a half-dozen shells before retiring "hastily with the cavalry by way of Grove Landing on the James and thence up the Beach road to Williamsburg." Shepard wrote in his report that he fired three rounds of spherical case before retiring. During the withdrawal, a Federal brigade of General William H. Emory was sent forward to cut off Shepard's section, then supported by Stuart on the Hampton Road. After attempting to thwart this threat, Stuart and Shepard's section were forced to retire to Grove's Landing near Carter's Grove Plantation. During the flight from the enemy, Private James W. Smith, "having been taken sick, was sent to the rear and was probably taken by the enemy."

While Shepard kept busy near Grove Landing, the bulk of Pelham's Battery moved from its position at Bigler's Wharf to join Stuart. The conditions were horrible. Due to the heavy rains of early May, the carriages and caissons trudged along through poor, muddy roads. Finally, on May 5, between 10 and 11 a.m., the battery received orders from Stuart to proceed immediately to Williamsburg. Despite the fact that Shepard's section had rejoined the battery, Pelham left five of his pieces in the city proper, before joining Stuart at 2 p.m. on the field near Fort Magruder with two 12-pounder howitzers and one 12-pounder Blakely rifle.

Having been near Fort Magruder since after dawn, Stuart had been actively monitoring enemy activity since General Joseph Hooker began the assault upon the fort with his division at about 6 a.m. Stuart, holding his cavalry near the two redoubts on the right of Fort Magruder, had "assumed the direction of the pieces within those redoubts during the rest of the day." Pelham's guns had arrived just in time to witness General John J. Peck's Brigade reinforce Hooker's advance on the fort.

Pelham's guns were an appealing sight to the gunners of the Richmond Fayette and James Dearing's Lynchburg batteries that had been heavily engaged since morning. Dearing and a section of the Fayette Artillery had just withdrawn into Fort Magruder when Pelham came onto the field. Taking position to the right and in front of Fort Magruder, the Stuart Horse

Artillery immediately opened fire on the enemy occupying the woods on the road to Lebanon Church. Dearing later reported that the fire of all of the guns present "was kept up incessantly by my two guns and one other in the fort, and by Captain Pelham's Horse Artillery outside, until the enemy ceased firing and were out of sight."

In the heat of the affair, Pelham detached 2nd Lieutenant William C. Elston with two men to bring off some captured artillery. In the attempt, all of the detachment lost their horses from enemy fire and in a few minutes were forced to return on foot. With the prizes lost to capture for the moment, Pelham continued to hold his gun position under heavy fire while General Daniel Harvey Hill's division moved up and deployed in front. While the infantry formed, Pelham was ordered by Stuart to remove his guns to the left, where they returned to battery on the Yorktown Road.

Stuart later reported of Pelham's battery that, "before the order could be given Pelham's battery was speaking to the enemy in thunder tones of defiance, its maiden day on the field, thus filling its function of unexpected arrival with instantaneous execution & sustaining in gallant style the fortunes of the day, keeping up a destructive fire upon the enemy until our infantry having reformed, rushed forward, masking the pieces."

Stuart later wrote his wife that he "consider[ed] the most brilliant feat of the 5th to have been the dash of the Stuart Horse Artillery to the front. Coming suddenly under a galling fire from the woods, from a reinforcement of the enemy, they wheeled into action sustaining in the most brilliant manner the fortunes of the day till the Inf'y could come to their support, and all the time under a continuous Inf'y fire of 200 yards or less distance." Stuart also saw fit to give the title "gallant" to his artillery commander in his report of the action at Williamsburg. It was not the only time that adjective would be used when referring to the young commander.

Moving next to a position on the Yorktown Road, the Stuart Horse Artillery opened an enfilade fire on the Telegraph Road. Pelham later reported: "Here the metal bed of the elevating screw of my Blakely gun gave way; but it was retained on the field and did good service. I remained in this position until 5 p.m. when I withdrew for want of ammunition.... I fired 286 rounds of spherical case and 4 of canister from the 12-pounder howitzers and 40 percussion shell and 30 solid shot from the Blakely gun." Finally retiring at 5 p.m. for want of ammunition, the three guns had expended a total of 360 rounds. Of the crews that served the guns that day, only three were wounded, including Summers and Gibson. Losses were heavy in horses: four had been killed, however, thirteen had escaped from horse holders. By the close of the day, all of the escaped horses but two had been recovered.

Continuing in his praise of the "Gallant Pelham" and his pet battery, Stuart wrote: "under the indefatigable exertions of Captain Pelham, ably seconded by his lieutenants and non-commissioned officers, they that day won the name of veterans. The daring and efficient conduct, under circumstances of great personal danger, of Capt. John Pelham, First Lieutenant James Breathed, William McGregor, Sergt. Charles R. Ford, Corpl. Moses Febry, gunner to the Blakely, and the brave men under them, came under my own observation."

As Johnston continued the retreat, Stuart's Brigade with the Stuart Horse Artillery moved to the west end of town until Tuesday morning. From there, Pelham reported, they moved and made a brief encampment at a brick school house, half a mile from Burnt Ordinary (now Toano).

At 10 a.m. on May 7, the battery was again called upon by Stuart to return to the rear. With a howitzer left at the Methodist Church about 1 ½ miles beyond Burnt Ordinary, Lieutenant Breathed held firm while a rifled gun was placed in position a mile farther on. The guns, opposing Colonel William W. Averell's two squadrons of the 8th Illinois and 3rd Pennsylvania Cavalry, made quick work of the enemy when they appeared at the edge of the woods in front. After the cannon fired only five well-placed shots, the cavalry scattered. Pelham then "retired in the rear of the main body of our forces."

Moving on to New Kent Court House on May 8, the Stuart Horse Artillery continued in the following days to engage in run-and-shoot encounters with the enemy. Finally halting their drive to Richmond near cavalry headquarters and about two miles east of Richmond, Pelham's company then moved out for a few days on the Valley Road, six miles from the city, and resumed drilling. The battery did not participate in the action at Seven Pines due to the poor condition of the roads.

While the brigade was idle, Stuart sent out John S. Mosby on a scouting expedition around the right flank of the enemy. Upon returning, Mosby reported that the enemy's right was protected only by a small screen of cavalry and offered an irresistible temptation for a raid by Stuart. Stuart had to make his case to a new army commander. With the wounding of Johnston, the army generalship had shifted to Robert E. Lee, someone with whom Stuart had been better acquainted in days past. On June 10, Stuart met with Lee and proposed a raid into McClellan's rear to disrupt his supply lines and communications, as well as attempt to ascertain further Federal intentions. Lee granted Stuart his expedition with only one additional caveat: that of "driving in his foraging parties, and securing...grain, cattle, & c., for ourselves." Stuart was to ascertain the disposition of Federal forces near the Virginia Central Railroad. While Stuart must have glowed with excitement at the prospect, Lee did caution him to not "hazard unnecessarily your command...be content to accomplish all the good you can with out feeling it necessary to obtain all that might be desired."

Taking only 1,200 men and horses that could withstand the endurance-trying expedition, Stuart gathered men from the 1st, 4th, and 9th Virginia Cavalry, the Jeff Davis Legion, and, from the Stuart Horse Artillery, a 12-pounder howitzer under Lieutenant Breathed and a rifled piece under Lieutenant McGregor. Pelham, with the rest of the battery, was to remain with the forces left to scout and picket for Lee's army. Moving his command to Kilby's Station, not far from Brook Church, Stuart held the secret of the mission to himself while he ordered his cavalry contingent to prepare three days' rations.

At 2 a.m. on June 12, Stuart was up and ordering his troopers to mount in ten minutes. Setting out initially to deceive the enemy, the march followed a route up the Brook Turnpike to near South Anna Bridge, seven miles from Hanover Court House. One trooper from the 1st Virginia Cavalry thought that the group was on its way to join "Stonewall" Jackson in the Shenandoah Valley. Bivouacking that night near Hanover Court House, the expedition continued before dawn the next day. It rode eastward and encountered elements of the 5th U.S. Cavalry near Gibson's Mill or Haw's Shop, where the famous Captain William Latane, of the 9th Virginia Cavalry, was killed. While elements of the cavalry were sent on various probes into the enemy, the horse artillery detachment appears to have remained as a "headquarters" element with Stuart, being reserved only for emergency use for most of the expedition. Prisoners, wagons, and supplies were bagged. What could not be taken was torched, leaving a trail of flames in their wake.

At one point on June 13, one of Breathed's two guns became stuck in a mud hole. Despite yelling at and thrashing the horses to pull it out, the gun refused to budge. As the situation became more frustrating, a sergeant quietly called the lieutenant aside with a suggestion. Agreeing to the thought, Breathed ordered up a small barrel of whiskey, recently "acquired" from enemy stores and placed it on the carriage. With the promise that the barrel would be liberated when the gun was out of the mud, the men made certain it was soon freed from the quagmire.

On the 14th, Stuart came upon the remains of the Forge Bridge. Finding the stream unfordable, Captain W. W. Blackford took men and began tearing down an old warehouse, using the timbers to construct a makeshift bridge that allowed the crossing. Finally returning to the safety of the Confederate lines, Breathed's section had only lost one limber in the ride around McClellan, when it was bogged in fording an offshoot of the Chickahominy.

While Breathed and McGregor had been on the expedition, Pelham may have been busy recruiting, as a large group of Marylanders swelled the ranks of the battery even more. Soon afterward, the Stuart Horse Artillery found it necessary to reorganize. As a result, Daniel Shanks, formerly of the 1st Maryland Infantry, was promoted to 3rd lieutenant and Corporal Moses A. Febry was promoted to 4th lieutenant. On June 25, the Stuart Horse Artillery had 180 horses. Stuart's Brigade also saw growth, now consisting of the 1st, 3rd, 4th, 5th, 9th, and 10th Virginia Cavalry, the Cobb (Georgia) Legion, the Jeff Davis Legion, a squadron of the Hampton Legion, three companies of the 1st North Carolina Cavalry, and the Stuart Horse Artillery.

On June 25, Stuart prepared the 1st, 4th, and 9th Virginia, Cobb's Georgia Legion, the Jeff Davis Legion, and Hampton's South Carolina Legion, along with the Stuart Horse Artillery, to be put on the road once again. At 8 a.m., camp was broken and the movement initiated to join "Stonewall" Jackson's troops arriving from the Shenandoah Valley. Stuart's Brigade was to shield Jackson's flanks while he struck McClellan's right flank. Bivouac was made that night near Ashland.

The following morning, the route was regained, moving through Ashland and Hanover Court House into New Kent County, seven miles from Cold Harbor. Finally meeting with Jackson, the 1st Virginia went forward to protect Jackson's infantry columns from surprise on their movement to Pole Green Church.

With Captain Charles R. Irving, of the 1st Virginia, deployed to seize the bridge over Totopotomoy Creek, Stuart met with resistance as the Federals had also begun to destroy the bridge. A courier hurried back to Pelham at once, and a couple of howitzers raced forward to a point near Doctor Shelton's and wheeled into battery. Soon the guns were barking out rounds of canister into the bridge-burners. Their is some contradiction in books relating to this event however; Milham cited Pelham as performing this, but Douglas Southall Freeman gave credit to Captain James Reilly's North Carolina Battery. Milham's personal notes, while not entirely available now, may have held the truth to the matter. Nevertheless, with the route saved, Jackson was able to cross the following day.

Continuing his efforts to protect Jackson's left flank, Stuart marched his men eastward on the morning of June 27. With the 4th Virginia, the Jeff Davis Legion and the Stuart Horse Artillery, the columns continued past Bethesda Church and Beulah Church. On the road to Old Church, Federal skirmishers were encountered and broken up with cavalry and artillery fire. Enemy cavalry skirmished at different points along the route before Stuart's gray columns halted for the evening near Old Cold Harbor. However, with the sound of battle clearly echoing to there front, there would be no rest for some men of the Stuart Horse Artillery.

Stuart reconnoitered with his staff and the Stuart Horse Artillery toward Old Cold Harbor Tavern. Upon their approach, General George Sykes' divisional artillery kept up a fearful fire to which Pelham's guns were called upon to respond soon after arriving near 5 p.m. With only a Napoleon and a Blakely deployed, Pelham continued a defiant fire against the guns of Tidball's & Weed's batteries on McGehee's Hill. Soon Pelham, as seemed to be the case throughout his career, was put to a further test when yet more Federal artillery arrived from Grapevine Bridge. Placed on a ridge away to one side of McGehee's Hill, the new eight pieces wheeled into battery and added the chimes of their guns to the duel. Well within view of the Federal artillery, Pelham's guns kept up a steady fire until one of the new enemy guns on the field found its mark and struck the carriage and trail of the Blakely. Now with only one gun to protest the uneven fight, Pelham refused to waiver and responded with more fire from the Napoleon. Concerned for his "adopted son," Stuart sent forward a squadron of cavalry, under Colonel Thomas R. R. Cobb, to Pelham's support. After finding the enemy's shells most uncomfortable, Cobb chose a safer spot from which to support the gun by moving back into the woods to avoid the shelling. Stuart and Jackson both observed the daring of the young artillery captain.

Clearly seeing the need to reinforce Pelham, Jackson, the old artillerists, sent in the batteries of Brockenbrough and Courtney and a section of Carrington's Charlottesville Artillery to aid the captain. These batteries advanced en echelon. According to John Esten Cooke's edited diary of the events, he was the one sent to order at least one battery to the left of the road to advance en echelon on the hill to the left of Pelham.

While not among the section of the Charlottesville Artillery that deployed near Pelham, Private Wilbur Fisk Davis of that battery recollected seeing one gun (Pelham's) deploy not far from their front and open on the Federal batteries. "It all came to a stop," declared the Tarheel-born author of the Charlottesville Artillery, "with the yelling charge which captured the [enemy] battery."

When a lull in the fighting finally offered the opportunity, Captain Pelham rode back to the position where Stuart, Jackson, and the staffs of the two officers had watched the artillery duel. When the young Pelham rode up among the staff, the stoic Stonewall looked upon the powder-dirtied and winded young man with appreciation, smiling and offering his hand in admiration and congratulation. Pelham reportedly responded by bowing deeply and blushing.

Only a day after the fight at Gaines' Mill, the Stuart Horse Artillery was with Stuart again on a raid on the York River Railroad, believed to be McClellan's most likely route of retreat. Advancing on Dispatch Station,

Stuart's troopers tore up the railroad and cut the telegraph wires before moving on toward Tunstall's Station. Near there a squadron of enemy cavalry, now on the far side of Black Creek, had destroyed the bridge. Likewise, Federal artillery was noted behind the Confederates on a height. Calling upon his own "long-arm," Stuart ordered Pelham up and with the howitzers opened on both positions. Stuart described the position as "a difficult stream, whose abrupt banks and miry bed presented a serious obstacle to our progress." The bloodless event over, the enemy guns on the heights were later discovered to be nothing more than "Quaker guns."

The following day the cavalry continued to the White House plain and found that numerous supplies had been burned in the retreat of the enemy. The White House too had been torched by Brigadier General Silas Casey's men, despite the fact that the residence was already a noted national landmark, having been the residence of the Widow Custis. George Washington had spent much time there, courting and marrying the widow at the house. It would be their home in the first few years of the marriage, and, in years that followed became the residence of General William Henry Fitzhugh ("Rooney") Lee. In the time preceding the Seven Days, Mrs. Robert E. Lee had stayed at the home and had to pass through Federal lines when McClellan arrived. Protected by McClellan's staff, Mrs. Lee made a last effort for the protection of the home prior to her departure. On the door she left a note that she had penned stating "NORTHERN SOLDIERS WHO PROFESS TO REVERENCE WASHINGTON, FORBEAR TO DESECRATE THE HOME OF HIS FIRST MARRIED LIFE, THE PROPERTY OF HIS WIFE, NOW OWNED BY HER DESCENDANTS. A GRAND-DAUGHTER OF MRS. WASHINGTON." General Casey claimed to have tried to prevent the firing of the home, but it was nevertheless torched by some Federal soldiers.

Further investigation near the landing found the U.S.S. *Marblehead* just lying in the river. Stuart set out sharpshooters from the 1st Virginia at once. Spread out along the riverbank, the men picked away at the gunboat, driving the enemy from the weather-decks and causing a general annoyance. "To save time" Stuart ordered up Pelham and a howitzer to fire on the vessel.

The *Marblehead* skipper, convinced that the exploding spherical case was not a good match, steamed away downstream. While the Confederate cavalry tended to the stores left behind by the enemy, Pelham pressed the pursuit of the gunboat until a long bend in the Pamunkey made it impossible to continue. The log from the *Marblehead* gives a clear picture what the morning was like from the other end of Pelham's gun. At 8:10 a.m.:

> ...fired 1 five second shell from X1-inch gun at... rebel cavalry seen on shore about 1,000 yards... 11:45 underway & steamed down the river, the rebel artillery firing at us

after we were underway, their shot going over us and some falling short. In hauling the boats alongside the second cutter capsized, losing all the oars, cushions, etc.

From White House landing, Stuart marched again on June 30, concentrating on Long Bridge. Moving with Fitz Lee's 1st Virginia, a howitzer deployed north of the Chickahominy between Bottom's Bridge and Providence Forge. Though the detachment encountered Federal pickets at Long Bridge, the howitzer's trail gave way, rendering it useless for further use. "Met" Coode was the only man of the battery recorded as wounded in action at White House.

While the howitzer was detached on its jaunt with the 1st Virginia, Pelham had gone with the 4th Virginia and the Jeff Davis Legion toward Forge Bridge. With only two serviceable howitzers remaining, Pelham's section saw action at 11 a.m. when he opened the guns at 400 yards, opposing an enemy artillery force. After an exchange of rounds, the two Federal rifled pieces withdrew to a line of hills on the far side of the river. The enemy cavalry and infantry also disappeared. Continuing to reconnoiter, Lieutenant Colonel Will T. Martin, in charge of the Confederate contingent, sent a small force across the river in the afternoon and again encountered a squadron of enemy cavalry. Having anticipated further action, Pelham was quickly out on the bridge with his two howitzers and ordered by Martin to shell the road up which the enemy was advancing. The rifled guns barked out their shots and the enemy replied with a salvo of their own, causing little effect upon Pelham's guns. After no more than fifteen minutes, the Federal cavalry broke up and retired from the field.

Moving into bivouac that night beside Forge Bridge, the men were roused early at 3:30 a.m., July 1. With orders to protect Lee's left on its move toward the James River in pursuit of McClellan, Stuart drove his men on to Bottom's Bridge. Arriving as Jackson's men were on the road, the cavalry was held up by movement of Jackson's men for about two hours before the brigade moved back to Forge Bridge. Stuart, advising Jackson that he would take the side-roads south of the Chickahominy to reach the left, pressed his horsemen onward in the hot and humid July sun. The heat took a tremendous toll on the horses of the horse artillery, compelling Pelham to change horses on the limbers and caissons often. Throughout the day the sounds of tremendous artillery barrages and musketry could be heard emanating from the fighting at Malvern Hill. Continuing into the night, the column made such a remarkable amount of noise that Stuart was convinced that it in itself played an important role in the retreat of the enemy. It was his opinion that the enemy were left to think, due to the noise, that they were in danger of being cut off.

Through a drenching rain during the morning of July 2, Stuart continued the pursuit of the enemy on a road littered with haversacks and numerous other accouterments. Some of Martin's Mississippians with one of Pelham's 12-pounder howitzers moved out on the road to Haxall's Landing and encountered the U.S.S. *Monitor* and the U.S.S. *Galena*.

That night, Stuart called upon Pelham to take a single howitzer with a squadron from the 1st Virginia Cavalry to move on the River Road, and if possible toward Westover to attempt to cut off the retreat if the enemy was using that route. As the party moved through the darkness, they encountered the retreating enemy between "Shirley" and "Westover," protected on the right by Herring Creek. Most of the Federal encampment spread out below Evelington Heights. At once Pelham returned a report to Stuart stating the entire disposition of the enemy and a proposal for more artillery to be planted on the heights. Upon receiving the report, Stuart dispatched a report to General Lee via Jackson before pressing on with his own brigade, less the 9th Virginia to cover the rear. After routing the small cavalry force occupying the position, Pelham's howitzer was ordered at 9 a.m. to open on the enemy camps, sending teamsters and horses flying through the camps below. The shelling kept up until 2 p.m., holding off a weak enemy resistance. Despite his constant communications to Lee and the word that Longstreet was en route, Stuart began to grow concerned that the opportunity was rapidly slipping away. General Franklin began to beef up the force against Stuart and had a battery moved east of Herring Creek. Soon the Confederates also felt the heat of artillery bearing down against them. Furthermore, under protection of the battery, Federal infantry gained a position on Stuart's right that made it even more uncomfortable. To make matters worse, Pelham's lone gun was reduced soon to two rounds of ammunition. Receiving word that Longstreet had lost his way and was at Nance's Shop, Stuart was forced to retire.

After the affair at Evelington Heights, Pelham's company went back into camp for a day before being called upon once again. During the night of July 5, the Blakely gun under Lieutenant McGregor, along with a section of A. L. Rogers' Loudoun Artillery and C. W. Squires' 1st Company, Washington (Louisiana) Artillery, was ordered to Colonel Stephen D. Lee. Lee's detachment, under orders from General R. E. Lee, was to proceed to the banks of the James River, above Charles City Court House. Following in the rear of the 5th Virginia Cavalry, the hefty contingent arrived after midnight on the farm of Thomas Wilcox near a point known as Wyanoke on the James River, some four miles below Charles City Court House. Moving into battery, the guns at 7 a.m. fired upon the transport *Juniata*. The transport immediately extinguished her lights and quickly struggled to free herself from the point where she had run aground. After some twenty shots, some striking below the waterline, the boat came about and made her way back down the river, throwing cargo overboard to expedite the escape. The transport limped away with two men wounded and the hull leaking badly. The firing had stirred up quite a bit of activity farther down the river among the gunboats as well.

Apparently, the *Juniata* soon came into contact with the U.S.S. *Monitor, Galena* and *Sebago* and told the gunboats of being fired upon by a battery of four guns. The *Monitor* and *Galena* proceeded upriver "for the scene of action, convoying the *Juniata* and tugboat *Red River*." Once the gunboats were abreast of the bluff, they fired three shells in the direction of the Confederate guns but received no response. Colonel Lee wrote, "Several gunboats of the enemy immediately came to the vicinity and commenced shelling the woods and fields, but without damage." Farther down the river, and in response to the sound of the cannonading, the U.S.S. *Mahaska* and U.S.S. *Delaware* prepared to make way also, though the *Delaware* ended up not making the journey.

Anchored about 1/4 mile from Fort Powhatan, the *Tiago* responded at 2:30 p.m. to the fire and went down river, shelling the position of the Confederate guns. By 3 p.m., the U.S.S. *Mahaska* had come to anchor at Westover. Colonel Lee reported that:

> About 5 p.m. six or eight transports (schooners), towed by a tug and conveyed by a gunboat, came up river, as also several small river steamers. The guns were drawn up behind a bluff at Wyanoke, and when the transports were opposite ran to the edge of the bluff and opened on them. Some twenty shots were fired into the transports, damaging them severely. The tug was also damaged. The gunboat opened on the battery as soon as discovered, firing shell, spherical case, and grape from her large guns. She fired some 30 rounds. No damage was done the battery, all her shot either falling short or passing over our heads.

Tiago's logs showed that it had fired 59 rounds of shell before coming to anchor off Windmill Point. By 7:30 p.m., the gunboat reported that it was within range of the battery. From 5-7 p.m., the gunboats shelled the Confederate camps.

The mission of harassing being well accomplished, Lee retired the guns to Charles City Court House and bivouacked for the night. On the morning of July 8, he was ordered to move down the Charles City Road in the direction of Bottom's Bridge, "as the enemy was advancing in large force with infantry and cavalry." The threat of advancing dissipated and the order to the direction of Bottom's Bridge was revoked; the guns returned to their respective camps. McGregor's Blakely had expended only 20 rounds during the expedition (the least rounds out of all guns involved).

On July 10, the camp of the Stuart Horse Artillery was moved to the Timberlake Farm near Atlee's Station in Hanover County, ten miles north of Richmond, leaving only a small detail on outpost before McClellan. Life was about to improve for a short period. Rest was well deserved following the harrowing schedule since the opening of the Peninsula Campaign for these now battle-seasoned veterans of the horse artillery. To add to the numerous spoils picked up along the roads in the past month, Pelham also received caissons and a couple of Napoleons, a twelve-pounder howitzer, and assorted equipage. Likewise, there was time to repair the older pieces and strengthen the ranks.

By July, Pelham's company had also begun to show losses in desertions. On July 16, Corporal Moses Febry, with a detachment of three men, left camp for Floyd County and the recovery of some of those deserters.

It wasn't until July 19 that Pelham's battery was disturbed by anything significant. Even then, the 150-man battery was only charged with participating in the review of Stuart's command. Just over a week since their arrival at Timberlake's, the men fell in and boldly showed their pride after a successful campaign. On July 29, Stuart held another review in which the Stuart Horse Artillery also participated.

Reorganization also took place during this time. With Stuart being appointed as major general of all cavalry forces in the Army of Northern Virginia, it was inevitable that the rank structure would shift upward for many. With promotions also came brigade reorganization. Wade Hampton was promoted to brigadier general commanding his own brigade, which included the 1st North Carolina, 2nd South Carolina, the Jeff Davis Legion, and the Cobb Legion. The 2nd brigade was commanded by Brigadier General Beverly H. Robertson and included the 2nd, 6th, 7th, and 12th Virginia Cavalry, and the 17th Virginia Battalion. Brigadier General Fitzhugh Lee's 3rd Brigade included those more familiar to the men of the Stuart Horse Artillery, including the 1st, 3rd, 4th, 5th, and 9th Virginia Cavalry. The artillery was also involved in reorganization to a point. Chew's Ashby Battery and James F. Hart's Washington (South Carolina) Artillery joined the fold of what was becoming the Stuart Horse Artillery Battalion. Despite the addition in batteries, and despite Stuart's efforts, Pelham remained captain.

Continuing a peaceful camp-life existence, with the usual drilling, the men of the Stuart Horse Artillery basked for a time in the glory of the recent months past. On August 4, Pelham was called upon with his battery to move with Fitz Lee's Brigade to reconnoiter Federal activities and possible river transports near Fredericksburg. Moving to Bowling Green and bivouacking that night, the first day passed without event. In the intense heat of the following day, the column of gray horseman pressed onward and turned toward Port Royal on the Rappahannock. In their trek, they caught up with a squadron of Federal cavalry that had sacked personal residences in Port Royal. Bagging the lot of men and the loot as well, the expedition continued toward Fredericksburg, the guns of the Stuart Horse Artillery still following.

Moving on to Fredericksburg, the column made a left turn after passing Moss Neck Creek and bivouacked that night near Grace, or Round Oaks, Church, 10 to 12 miles below Fredericksburg. On the morning of August 6, Stuart began receiving reports of a strong Yankee column on Telegraph Road. Proceeding farther, the Confederates encountered a large force of infantry, cavalry, artillery, and baggage under Brigadier Generals John P. Hatch and John Gibbon moving toward Hanover Junction. It appeared that the enemy objective was the Fredericksburg and Virginia Central Railroad. Stuart raced Fitz Lee's Brigade and the Stuart Horse Artillery behind the 4th Virginia Cavalry over the cross-road that ran into Telegraph Road at Massaponax Church. The baggage train was a tempting morsel for Stuart and it had to be taken. Dashing across the rolling hills, the cavalry caught up with the enemy rear and rushed in, capturing the train and escort.

After bagging the lot, Stuart redeployed his men. Leaving a reserve of half of his cavalry and artillery to guard the train, Stuart moved on with the rest of the force to bear down on the enemy column ahead. Catching them at the Po River, Pelham's guns were quickly rushed forward.

David Cardwell later recollected what one horse battery member recalled what it was like to see a horse battery race into action and might have closely resembled the action of the warm summer day:

> Down the crowded highway galloped a battery.... The field fence is scattered while you could count thirty, and the guns rush for the hills.... six horses to a piece, three riders to each gun, over dry ditches where a farmer would not drive a wagon, through clumps of bushes, over logs a foot thick, every horse on the gallop, every rider lashing his team and yelling.... The guns jump two feet high as the heavy wheels strike rock or a log, but not a horse slackens his pace, the cannoneer leaning forward in his saddle. Six guns, six caissons, six horses each, eighty men Race for the brow of the hill, as if he who reached it first was to be knighted. A moment ago the battery was a Confused mob; we look again, and the six guns are in position, the detached horses hurrying away, the ammunition chest open.... We have scarcely obeyed when "Boom, boom!" opens the battery, and jets of fire jump down and scorch the green trees under which we fought and dispaired. What grim, cool fellows those cannoneers are! Every man is a perfect

> machine. Bullets splash the dust in their faces, but they do not wince; bullets sing over and around them, but they do not dodge. There goes one to the earth, shot through the head as he sponged the gun. The machinery loses just one beat, misses just one cog in the wheel, and then works away again as before. Every gun is using short - fuse shells. The ground shakes and trembles. The roar shuts out all sounds from a battle line three miles long, and the shells go shrieking into the swamp to cut trees off short, to mow men until their corpses cannot be recognized as human.
>
> You would think a tornado was howling through the forest, followed by billows of fire, and yet men live through it.... Now the shells are changed to canister, and the guns are served so fast that all reports blend into one mighty roar. The shriek of a shell is the wickedest sound in war, but nothing makes the flesh crawl like the demoniac, singing, purring, whistling grapeshot and the serpentlike hiss of canister. Men's legs and arms are not shot through, but torn off; heads are torn from bodies and bodies are cut in two.

Placed in an elevated position while the cavalry charged on ahead, Pelham opened with his first rounds, limbered, moved ahead, unlimbered and fired again, repeating the horse artillery specialty along with the cavalry charge.

As the Federal infantry began to form in battleline, Stuart became more concerned about what was taking place on his flank. While descending upon the rearguard, Stuart had detected movement to the south; Hatch and Gibbon had come about and were preparing to meet Stuart outright. Withdrawing to a position just out of reach, Stuart kept his cavalry and horse artillery in line as the enemy advanced. In a reverse leap-frog maneuver, Stuart lured the enemy along with him and away from Hanover Junction. Pelham's horse artillery followed suit and fell back from point to point while sustaining a hot fire on the Federals. In the withdrawal, the Blakely's mechanism broke, leaving only the howitzer to keep up a good fire. Having successfully bagged the enemy train and sufficiently thrown off a potential raid, Stuart was content to retire from the field and began to make his way back to Bowling Green, where the victorious cavaliers bivouacked for the night.

On August 7, Stuart moved his long line of cavalry back to Hanover Court House, carrying with him the train and 85 prisoners. Recalling the day, George Shreve wrote that it was "hot and many of our horses succumbed to the heat, and we impressed from the farmers to replace them.

Not far from Guinea station, we overtook a carriage containing two ladies with coachman, a fine turnout, drawn by a splendid pair of horses. We were compelled to appropriate these latter, very much to the disgust of the ladies, who protested quite vehemently."

Upon the return to Hanover, Stuart also received the news that Pelham's commission to major had been approved. Though it would not actually be final for another six weeks, Stuart had raised his "chief of artillery" to a position worthy of battalion commander.

With few days of rest between their raid on Gibbon's rear, the Stuart Horse Artillery began to recognize that there would be no break like the time after the Seven Days. On August 15, Stuart received orders to move to the Rapidan and Gordonsville to protect that portion of the Virginia Central Railroad. Starting at daybreak on the 16th, the columns of men moved to camp that night near Davenport's Ford on the North Anna River. From there on the following day, Fitz Lee was ordered by Stuart, though clearly not emphasizing speed, to move via Verdiersville for Raccoon Ford. Unknowingly, Lee's movement would take him the long way to Raccoon Ford, instead of the way that Stuart would have preferred. The next day at Louisa Court House, rations were replenished before pressing northward. While arriving ahead of, and waiting for Lee, Stuart held up at Verdiersville for the night and was nearly captured the following day. The episode of losing his plumed hat would be something Stuart would have to avenge personally.

Lee, much to the frustration of Stuart, finally arrived on the night of August 18. The long march on poor roads had been taxing and had worn out men and horses alike. Taking this into consideration, Stuart allowed a day to rest before moving on any farther.

On August 20 the march was renewed. With General Longstreet's leading columns, Fitz Lee's Brigade and the Stuart Horse Artillery moved across Raccoon Ford on the Rapidan well before dawn. Guns and caissons were particularly hard to get across due to depth of the ford. Once they had made it to the other side, the battery was quickly called upon for fire-power. Contesting the crossing of one of Stuart's brigades at another ford was a force of cavalry that held back the advance for some time. As Pelham's guns pitched in, the enemy began to break and withdraw. For Pelham, this wasn't enough. The running fight continued for more than eight miles, driving the enemy back to Kelly's Ford. Ultimately the enemy was pressed across to the other side of the Rappahannock.

Meanwhile, near Brandy Station, Stuart and Robertson's Brigade had their own fight with the enemy. Leaving a section of guns and two regiments to hold General John Buford at Kelly's Ford, the rest of Fitz Lee's

Brigade moved off to reinforce Stuart and Robertson. Learning from prisoners that General George D. Bayard was waiting for reinforcements to counterattack, Stuart began to create a defense. Pushing Fitz Lee's men into line with Robertson's on a position near Brandy Station, and posting Pelham's section on the right, Stuart now sat back and waited. The counterattack never came. Shreve later wrote: "All of our guns were in action, two Howitzers, being pressed forward, close to the enemy. We did effective work, and drove the opposing column before us... We had one man killed, a Alabamian. We then pushed on, and crossed the Upper Rappahannock."

Receiving orders to attack General John Pope's supply lines, Stuart started early on the morning of August 22. A mile or two above Beverly at Welford's Ford and at Freeman's Ford, Stuart struck the enemy again. At Freeman's Ford, Pelham was sent in with four guns, including some from Chew's battery, against Federal artillery and infantry on the other side. Dueling for hours, the enemy refused to yield. Forced to retire, Pelham shifted his guns up and down the Rappahannock to whatever ford justified the need for artillery support for the next few days. At one point in the activities of the day, both Fitz Lee and Pelham personally dismounted and aided in the operation of one of Chew's guns, the gunners being exhausted.

On August 24, Stuart moved in the afternoon to Waterloo Bridge with the Stuart Horse Artillery and Chew's Ashby battery. With a detail from the cavalry and a detachment of Longstreet's men the goal was to thwart any Federal effort made at destroying the bridge. Jackson's movements had to be shielded before the bridge was lost. One gun each from Pelham's and Chew's batteries, were engaged with the enemy battery on the opposite side. Gunner George Neese, of Chew's battery, recollected that the "Yankee artillery must be getting cross as they fired solid shot at us, trying to break our guns..." Holding all attempts at bay, the contingent was relieved from its important mission on Monday, August 25.

At 1 a.m. on August 26, the camp of the Stuart Horse Artillery came to life when the batteries prepared to accompany Fitz Lee and Robertson up the Rappahannock. On the road by 2 a.m., the strong cavalry force moved to Hinson's Mill. Catching up with Jackson at Salem, the brigades turned south. Stuart spread out his force to continue to cloak Jackson. Now separated from Lee and the main body of the Army of Northern Virginia, Jackson and Stuart were vulnerable to the heavy forces under General John Pope. During the afternoon, Pelham moved on toward Centreville with a few cavalry squadrons.

On August 27, orders came from Stuart to move a section of guns to the Orange and Alexandria Railroad to hold off a possible approach by the enemy. Having just been alerted by the train raids at Bristoe Station, the Federals were certain to have reinforcements sent out to reconnoiter or brush off any enemy force. Just as had been anticipated, just a few hours after sunrise a train approached and halted to unload its cargo on the other side of Bull Run. Two regiments of Federal infantry under Brigadier General George W. Taylor moved across the railroad bridge. Offering the enemy a chance to surrender, Jackson was ignored and the New Jersey brigade pressed onward. Joining with the Virginia batteries of Poague and Carpenter, Pelham placed his guns. Soon the unsuspecting enemy regiments were sent to cover by a massive artillery bombardment, running for the cover of the bridge. The Confederate infantry hurried in pursuit, while Pelham limbered up and joined the race. Other batteries present performed feats unknown to them on the field before.

Again on a hill near the tracks, Pelham's gunners went into battery and opened hard on the retreating foe. While the Federals fled pell-mell in every direction, General Taylor was struck by a fragment and mortally wounded. Enemy infantry and a later-arriving group of cavalry were pursued as far as Burke's Station by Fitz Lee. Meanwhile Pelham was ordered to do something of greater importance -- the securing of ammunition from the Federal stores at Manassas Junction.

Upon arriving at the junction, the men of the Stuart Horse Artillery took note of General A. P. Hill's men scavenging the goods left behind by the enemy. While looking for good ordnance supplies, they came upon two brand new 3-inch rifles. Without hesitation the new pieces were claimed. At once, orders were issued to hide two of the older howitzers in the woods while the limbers were to be brought up thereafter and hitched to the two new guns. Pelham's attention to the importance of ordnance over delicacies would not be overlooked by Stuart. Seeing to the other needs of the Stuart Horse Artillery, the general sat astride a sutler's wagon drawn by four excellent bay horses for the company. Ultimately, the company would appreciate the fine contents including oranges, lemons, wine, cigars and other items. The horses were fitted and used to draw one of the new 3-inch rifles.

With the busy undertakings of the day behind them, the company was put on the move once again at dusk. Apart from the detachments deployed to feel out the enemy, the rest of Stuart's force remained in the area of Manassas. Before leaving the junction, however, men put the torch to whatever could not be taken. In support of this, Pelham left a section at Manassas Junction. Meanwhile a second section had been deployed forward with A. P. Hill's division, eventually concentrating the remainder of the Stuart Horse Artillery near Centreville by Thursday morning. From that point, Pelham was ordered in support of Jackson's feint on Centreville.

On August 28, Pelham moved by way of Lewis' Ford through Bull Run. The Stuart Horse Artillery remained off to the side of A. P. Hill's division along the Warrenton Turnpike. At the ford, the horse artillery received a surprise from Federal cavalry, but were relieved by Lieutenant Breathed's lone piece acting as a rearguard. "A few well directed shots from Breathed's gun drove the enemy off." Breathed made quick work of the enemy and dispersed them. Continuing up the Warrenton pike, Pelham's Battery passed the Robinson house and overtook the rear of A. P. Hill's division at the intersection of the turnpike and the Manassas-Sudley Springs road early in the afternoon. Having an excellent eye for ground, the young major-select placed his pieces in a field just off the Sudley Road at the edge of an unfinished railroad grade. Jackson, in turn, ordered all artillery parked in the field.

Sometime in the day's movements, at least one man became separated from the Stuart Horse Artillery. While scouting the area of Warrenton with 150 cavalry, Federal Major Samuel H. Hairston caught one unnamed deserter from the Stuart Horse Artillery.

Despite being posted in the field with the rest of Jackson's artillery, Pelham could not be held down when the word came that twenty pieces were to be moved rapidly to the front. Shreve commented: "Capt. Pelham, seeing that a battle was imminent, and fearing he would have no part in it, obtain[ed] from General Stuart permission to report to General Jackson. He accordingly led the Battery at a gallop nearly the whole length of Jackson's lines, when observing the latter advancing, he dashed ahead of them, even beyond the firing line, and ordered, 'In Battery.'" Pelham reported:

> I took three pieces at a gallop through a thick woods in front of this general park, bearing to the right of the troops in position (Ewell's division). I crossed the old railroad about 1 mile from Groveton and took position between it and the turnpike. I neglected to state that one of my guns was unable to keep up and was lost from the battery, it being dark and the road narrow and winding. I reported to General Jackson, and he told me his chief of artillery, Major Shumaker, would show me a position. He conducted me across the railroad, as above stated. I moved on at a gallop until a heavy volley of musketry apprised me of the enemy's presence. I immediately put my guns in position and engaged them at about 50 or 60 yards. We continued the fight for an hour or more, when, our re-enforcements coming up, we drove the enemy back. During the latter part of this fight I had but one gun, the other having been taken off by the order of some mounted officer (it was dark and no one could tell who), while my attention was wholly directed to the right piece.

Actually, and as later reported in his post-action report dated in March, Pelham reported that after a half-hour, Shumaker ordered Pelham to fall back. However, "owing to the pole of one of my guns being broken I could not obey the order, and continued firing...." The two newly captured 3-inch guns had proved quite effective against the 19th Indiana Infantry. The firepower gained so much notice as to warrant General Solomon Meredith's attention, who, in turn, deployed two infantry companies to flush out the battery. Eventually, as the infantrymen crept forward and opened up through the grass, Pelham's guns were forced to retire. High plaudits were awarded from Pelham to Lieutenant Mathias W. Henry for his service that day.

Early on August 29, while Stuart was riding toward Groveton, he found that enemy sharpshooters had penetrated the woods and were moving toward the ambulances and trains. Threatened at being cut off, near 10 a.m. Stuart ordered Pelham to go into position on the east bank of Bull Run, just north of Catharpin Run, and open on the enemy in the woods. Pelham was to gather up all stragglers around the train and drive back the enemy. As the quartermaster was ordered to move the train toward Aldie, Pelham went into battery and hammered away at the enemy for an hour until two other batteries were ordered to the same position. Meanwhile, Pelham was sent with the battery to the right of the line and ordered to report to Jackson as Stuart went toward Hay Market. Arriving near Jackson, Pelham reported to the old artillerist who rode with him over the field, directing him to fight the guns "where fitting opportunity should occur." Pelham gave details in his report:

> Shortly after a courier reported that the enemy were falling back, and that General A. P. Hill wanted the artillery to press forward. I moved toward Groveton, and saw two batteries coming into position to play on the enemy's artillery near the town. I passed these batteries about 200 yards, and took position on the point of a ridge and opened upon their artillery. The position was held for nearly two hours, when the ammunition from all my guns except one was expended. The three batteries that were supporting me retired about the same time, and I was left alone, with one gun, exposed to the fire of a long line of batteries with a direct and flank fire. I dispatched Sergeant Hoxton to General A. P. Hill to inform him of my condition and ask him to send re-enforcements. After we had continued this unequal contest for fifteen or twenty minutes Sergeant Hoxton returned and reported that he could not find General Hill. I

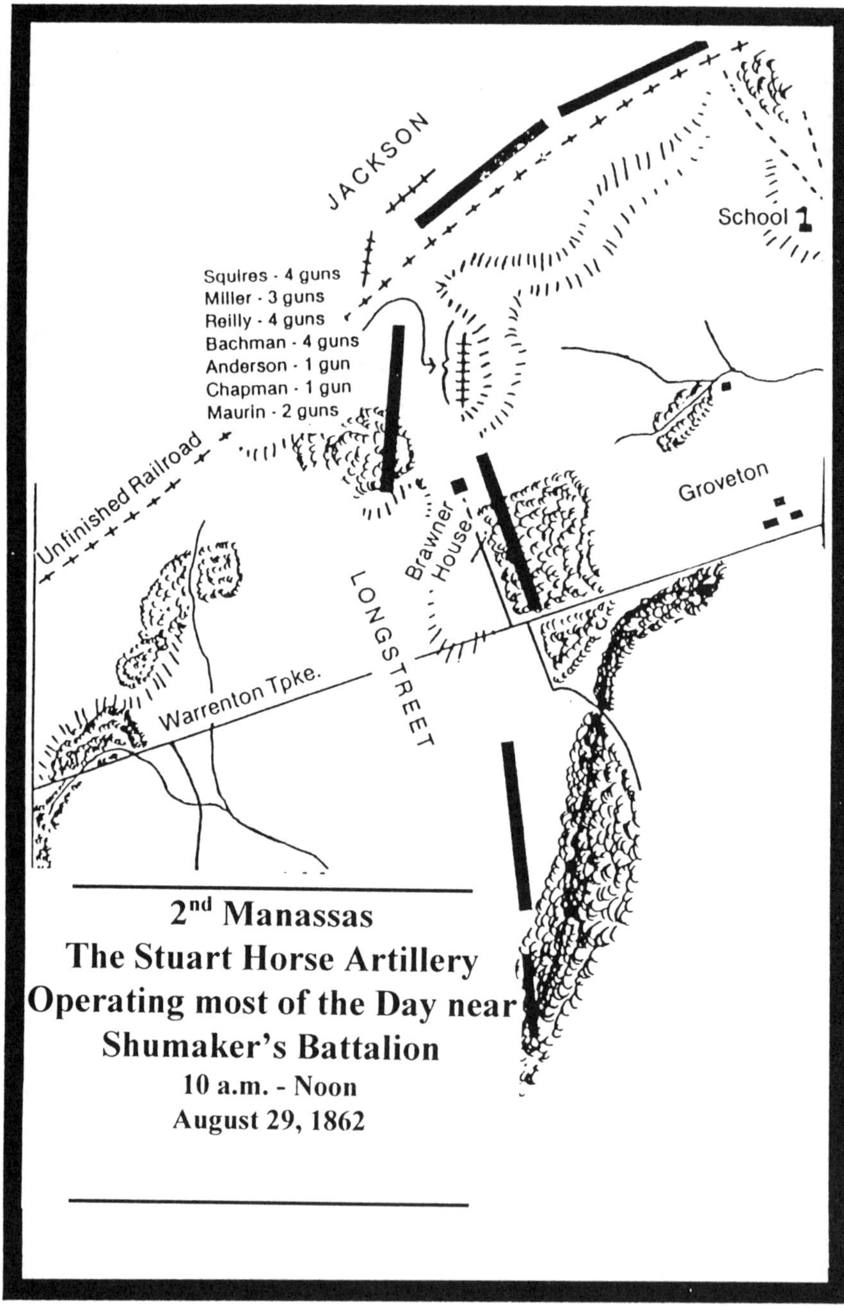

2nd Manassas
The Stuart Horse Artillery
Operating most of the Day near
Shumaker's Battalion
10 a.m. - Noon
August 29, 1862

then determined to retire, not, however, until the trail of my only gun had been struck and shivered. The accuracy with which my guns were fired and the rapidity with which they were served during both days was very gratifying, and the execution they wrought was very great.

Pelham gave high praise to several men in his battery including Breathed, McGregor, Sergeants R. T. Burwell and W. S. Dabney, and Sergeant W. H. P. Turner who was killed. Dabney was particularly cited in that he "acted admirably when left alone with his gun to fight at least twenty.... He fired his gun with the same precision and accuracy as before." In all, only one man was killed and five wounded. Among the wounded were Lieutenant Hoxton, Edward McCaffrey, John Arnold, and Charles A. Evans. Evans would have his right arm amputated as a result of the wound. Ammunition resupply would be particularly hard to accomplish after the battle. However, for the duration, the Stuart Horse Artillery would be held in reserve by Jackson on August 30 in the event of an "emergency."

On August 31, Pelham's guns were finally called upon once more and ordered to harass the enemy. Shreve recalled that "in the midst of a heavy thunderstorm" on September 1 at Ox Hill near Chantilly the guns also became briefly engaged. The enemy was in retreat and needed to be punished. Pelham was obliged to perform the service. That night, the battery moved with Stuart toward Fairfax Court House.

On the following day, Hampton's Brigade and the Stuart Horse Artillery were ordered to attack the enemy at Flint Hill. Once near the Federal position, Hampton encountered fire from both artillery and sharpshooters. In an effort to run them off the field, Pelham's guns opened and the enemy began to retire. In pursuit, Hampton's Brigade and Pelham's Battery followed close behind, soon overtaking them. The rifled piece opened and performed "with effect, scattering them in every direction." Bringing the cavalry up, Hampton pursued the Federals into the dusk, when it was deemed appropriate to continue "slow and cautious." Having moved forward without the support of artillery, Hampton found it necessary to withdraw when he came upon enemy infantry and artillery fire.

Following the expedition to Flint Hill, preparations were made for demonstrations upon Falls Church. On September 3, Stuart moved toward Dranesville, taking a few prisoners before pressing on to Leesburg. Pelham's accompanying guns unlimbered on Barrett's Hill, overlooking Falls Church, the and dropped a few shells into the place before the close of activities that day. On September 5, Stuart marched his brigade on to Leesburg and across the Potomac at White's Ford in the afternoon. From there, the battery, still attached to Fitz Lee, followed Stuart to Poolesville and halted for the night.

On September 6, while Fitz Lee made headquarters at New Market, near Urbanna, Pelham put in an urgent request and received 28 horses to resupply his losses from 2nd Manassas. Shreve recalled that sometime after the crossing, "At a point on the B&O Railroad, a lot of whiskey and salt fish fell into our hands. Rubber buckets and camp kettles were filled, and hung onto the gun carriages, and the contents sloshed out as we proceeded. The men made a grotesque appearance as they munched the fish and washed it down with the beverage, and some of them imbibing too freely were in a hilarious mood, and felt able to aniholate the enemy at once."

Undoubtedly, by the time the Stuart Horse Artillery made it into Maryland, and after five months of fighting, some traditions and practices must have become commonplace in the company. George Shreve wrote of group messes that joined together to obtain food and provisions. While on the march, men entertained themselves as they could:

> Sometimes a mess would entertain another mess at dinner with good cheer, war stories and song.... Frequently, while on the march, at night, one detachment, would strike up some inspiring song and the refrain would be caught up by the other detachments, all along the line of march, or around a Camp fire, the whole Company would gather by some impulse, and indulge in songs, relate stories until far into the night.... Playing cards was also a great pastime, mostly indulged in for the pleasure, but by some for the winnings.

Shreve also recalled that upon entering Maryland, "Gen. Lee issued stringent orders against taking any property by the soldiers. Some of the men came into camp one morning with a pig, and declared the pig had attacked them, and they were obliged to kill it in self defence. It was keenly enjoyed for breakfast and no questions asked."

Remaining here until September 10, the staff and officers of Stuart's cavalry and artillery partook of a fanciful ball. Major Heros Von Borcke had found an unused female academy hall had it decorated with roses and battle flags, and with music supplied by the band of the 18th Mississippi of Barksdale's Brigade, had quite an event on the night of September 8. Sometime during the night, however, Federals drove in a cavalry outpost that forced the ball to break up. Sending out the necessary force to throw the enemy back, including Hart's South Carolina Battery, Stuart and the officers returned to the place and continued the ball until the wounded came in and pressed the ladies into service as temporary nurses.

In the days following the activity in the area of Urbanna, New Market, and Liberty, Fitz's cavalry was involved in minor skirmishes with the enemy near Hyattstown before moving on to Frederick. Holding the town until the army made its way through, Fitz was again challenged on September 12, when word arrived that the enemy was advancing in force on the National Road. Stuart, realizing that two squadrons were on picket near the Monacacy Bridge, needed to protect them long enough to allow their withdrawal to the main cavalry force. Sending one rifled piece of Pelham's Battery to the assistance of the two guns already on the pike, Stuart then placed a squadron from the 2nd South Carolina to support the battery. The squadrons were able to come off their picket without loss. Meanwhile, Hampton began to withdraw to the city, "sending my guns [possibly Hart's Battery] to occupy a position which would command the road from the city to the foot of the mountain." Charging the enemy in the town, Hampton succeeded in capturing one gun and ten prisoners.

On September 13, Colonel Thomas Lafayette Rosser, an old West Point classmate and best friend of John Pelham, commanding a detachment of his 5th Virginia Cavalry, set out as ordered with the Stuart Horse Artillery to occupy Fox's Gap. Sometime during the passage of Fitz Lee's Brigade through Turner's Gap, Captain Pelham became literally lost. Officers and men alike began to grow seriously concerned until he showed up on September 14. No reason was ever documented for his temporary absence.

On September 14, Rosser was in position with his sharpshooters and artillery awaiting an attack by the enemy. When the attack finally came, General D. H. Hill sent three infantry brigades to Rosser's assistance. General Roswell Ripley later recalled that the enemy attack: "...was thwarted by two pieces of artillery belonging to Colonel Rosser's cavalry, which was speedily placed in position in our Rear on the Braddock road. A few well-directed shot and shell drove the enemy up the hill, leaving the road in our possession."

With most of Fitz Lee's Brigade, Pelham withdrew his artillery across South Mountain in the early morning of September 15. Acting as a relief for the infantry and a rear guard, Fitz Lee wrote that the brigade "was alone between the Federals and Lee at Sharpsburg to dispute their advance every foot of ground between the base of the mountain and Boonsboro. This was done with artillery, dismounted cavalry, and charges of mounted squadrons. The object having been accomplished, the brigade was slowly withdrawn and placed on the left of the line of battle at Sharpsburg."

When arriving at Sharpsburg, Pelham posted his artillery on Nicodemus Heights, the high ridge which gave considerable vantage over the northern part of the field and somewhat oblique to the Confederate line. Supposedly, on the night of September 16, while Stuart was walking about the position of the artillery and seeking Pelham, someone told him he was asleep in a fence corner. Surprised at hearing this, Stuart found the young soon-to-be-major and roused the young man stating "My dear fellow, don't you know that the cornfield is full of Yankees?" The men were soon at their pieces.

As the early morning of September 17 began to unfold, "Stonewall" Jackson rerouted every gun he could to Nicodemus Heights, being a former artillerist himself and realizing the good ground that Pelham had occupied. Among the batteries to fall in under Pelham were the Virginia batteries of the Bedford Artillery, Hampden, Alleghany, Danville, and Rockbridge Artillery. Additionally, the Louisiana Guard Artillery also took the heights. All 19 guns were to act under the orders of Pelham this day. Meanwhile a good amount of support was deployed to assist Pelham as necessary. Jubal Early was posted with his brigade nearby; the 1st Virginia Cavalry had dismounted and posted as well. Apparently, the 13th Virginia Cavalry also chipped in a fair amount of support to the "Gallant Pelham."

Captain Samuel D. Buck of the 13th Virginia recounted:

> We charged, skirmished and then went to work assisting the horses to pull the artillery in position.... Major Pelham was trying to get two Napoleon guns on a hill commanding their position but the horses could not pull them over the plowed field fields and up the hill so every fellow regardless of the shot and shell, took hold and almost carried both pieces into position and therein front of a corps of the enemy the Gallant Pelham fought those pieces, aiming and firing each piece as fast as the men could load, while our men supported him and poured the lead into their flank.

As the morning progressed, Pelham's mass of guns served well. Between Hooker's and Mansfield's assaults, Pelham shifted the guns of the Stuart Horse Artillery southward to Hauser's Ridge, behind the West Woods and swept the ground to the Dunker Church.

About 9 a.m., General Edwin V. Sumner's II Corps moved up. General John Sedgwick's lead brigade under General Willis A. Gorman moved into the West Woods and tasted the bitter pill from Pelham's guns. As Jackson deployed Early, Sedgwick gained the woods but was met and thrown back with more than 2,200 casualties when Major General John G. Walker's division and McLaws' division arrived from Harper's Ferry. Pelham's horse artillery continued to assist throughout the day, leapfrogging northward from Hauser's Ridge, keeping up with the infantry advance in the repulse. As McLaws' division attempted to pursue the enemy, however, Federal artillery in the North Woods ground the pursuit to a halt and costs the Confederates 40% casualties. Pelham was also severely punished for his support of the infantry in their advance; losses in horses were tremendous.

As the day progressed into the afternoon, Jackson considered making an attempt the following day on turning the enemy flank. In checking on the feasibility of this, Pelham made the reconnaissance while 5,500 men prepared for the opportunity including Stuart's cavalry and the Stuart Horse Artillery. Upon Pelham's observation, however, the attack was deemed ill-advised and considered extremely dangerous against the mass of Federal artillery found by Pelham.

The Stuart Horse Artillery had performed brilliantly that day under Pelham. His quick judgement in the placing of guns and the shift to Hauser's Ridge proved invaluable to the Confederacy that day.

As the Army of Northern Virginia began to retire on September 18, efforts were concentrated on getting back across the Potomac safely. On September 19, the Stuart Horse Artillery became involved in a defensive effort at Boteler's Ford until ammunition was exhausted. The company rejoined Stuart on September 20.

Following the heavy service of the battery at Manassas and Sharpsburg, Pelham made an unusual special comment pertaining to the service of Private Thomas Clifford Dudley. Dudley had entered the Virginia Military Institute in 1861, but had left to join the army. Of Dudley, Pelham wrote: "In two days of fighting at Groveton Heights and at Sharpsburg, South Mountain and various heavy skirmishes, he was conspicuously gallant. He is attentive... eminently qualified to command and discipline a company." Dudley would later transfer to the 5th Virginia Cavalry. Pelham's praise would not fall on deaf ears and by March, 1864, Dudley received a commission as 2nd lieutenant of infantry. Returning to the 5th Virginia Cavalry, however, Dudley would later be seriously wounded at Trevilian's Station in 1864, and die as a result of those wounds.

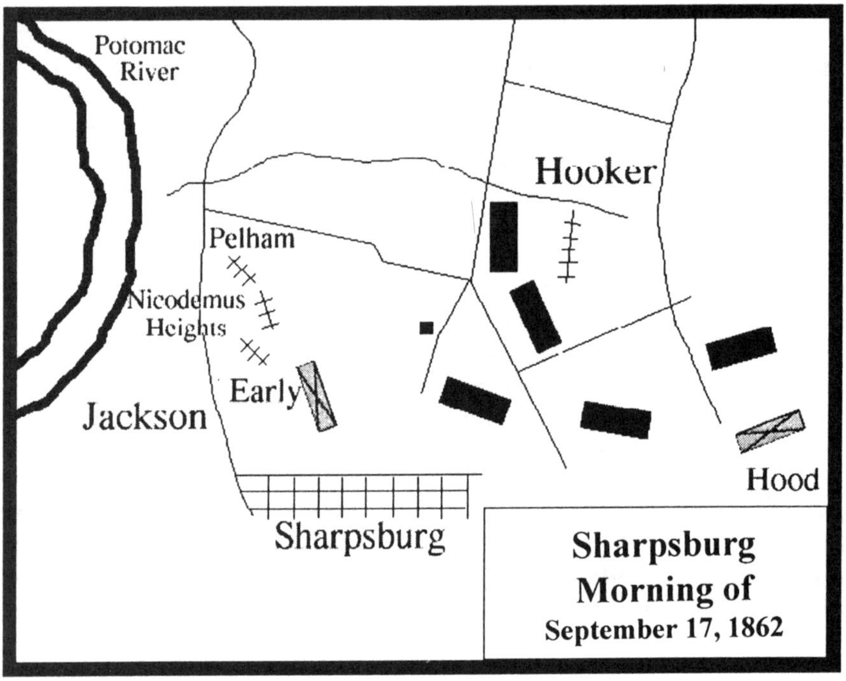
Potomac
River
Hooker
Pelham
Nicodemus
Heights
Early
Jackson
Hood
Sharpsburg
Sharpsburg
Morning of
September 17, 1862

Chapter 3

Chambersburg, the "Ten Days" & the Split of the Stuart Horse Artillery Battery
Fall & Winter
1862

Following Sharpsburg, the Stuart Horse Artillery was yet again involved in a major reorganization. With Pelham officially receiving his commission as major, the Stuart Horse Artillery Battery of six guns was divided into two batteries of four guns each. Breathed became captain of the 1st Stuart Horse Artillery and Mathias Winston Henry was made captain of the 2nd Stuart Horse Artillery.

Breathed, a familiar face in the ranks of the Stuart Horse Artillery, was remembered by H. H. Mathews in 1908 for his difference from Pelham in fighting the battery. In writing to Colonel A. Berkeley, Matthews stated that "They [Pelham and Breathed] were both fearless and courageous to a very high degree. Pelham was dashing and, at the same time, cautious. Breathed was reckless and not at all cautious -- would take his battery in places that would make your hair stand on end, depending on the bull dog fighting qualities of his men to bring him out safely."

Henry, a Kentucky native and Pelham's West Point classmate, had remained at West Point and graduated second to the last man in the class. Resigning his commission due to poor health, Henry made his way south in August 1861 and was commissioned as a lieutenant of cavalry. In less than a year of his promotion in the Stuart Horse Artillery, Henry would see his star rise even farther with promotion as a battalion commander in General James Longstreet's Corps. With Chew's Ashby Artillery and Hart's Washington (South Carolina) Battery, Pelham's Battalion was now at four full-strength batteries. Additionally, though not officially added until November, Marcellus N. Moorman's Lynchburg Beauregard Rifles would make the fifth battery under Pelham's Stuart Horse Artillery Battalion.

Much like the days after the Seven Days around Richmond, there was again time to relax following the return to Virginia. There was time to visit families and acquaintances. While visiting the estate of Major George H. Burwell at "Carter Hall," there was time for some quick recruiting as well. From "Carter Hall" that September, two sons joined the ranks of Henry's 2nd Stuart Horse Battery. Robert was enrolled as a lieutenant, while his brother George enlisted in the ranks. Other lieutenants to be raised in the shifting command of late September were Elston and Shepherd, both as 1st lieutenants.

Shreve recalled that the father of the two boys thought much of the battery.

> A unique and interesting character amongst us was "Major Burwell." His title was probably derived from the Militia service... During the Campaigns of 1862, the Major was with us as a sort of father or general benefactor of the Company. He was a man of high social standing and of means, living on his estate at Millwood, Clarke County, Va. In whatever section we happened to be located, he seemed to have friends of ample means, who would furnish him with provisions, including delicacies, which he would bring in to camp on his noble charger, and distribute to us bountifully. He was very proud of the deeds of the Battery, and did not hesitate to go into action with it on occasions. He was with us about a year, and visited us afterwards. His younger son, George, came to us later on; I think after our first campaign in Maryland. He was a high spirited boy, educated and of pleasing appearance, and good physique. On one occasion, I have forgotten the place, when a cavalry regiment was passing our battery in a charge, our guns being just then silent, he obtained permission to accompany it, and dashing on with it, he captured a cavalryman, with a fine horse and equipments, and displayed the trophies with great pride. I think he could not have been much over sixteen years old... After the surrender, he would not return home; but crossed the Mississippi, and went to Mexico, where he offered his services to Maximilian. On account of his youth, he was denied a commission at first; but meeting Queen Carlotta, she was so impressed with his appearance and enthusiasm that she championed his suit, and obtained it for him. He was made Captain, and in battle with the Mexicans, he was killed in a charge.

During the intermission Pelham had an opportunity himself to enjoy time on the Opequon at the "Bower" near Keedysville. At that home of Colonel Adam Stephen Dandridge in Jefferson County, Virginia, Stuart and his staff spent several days being warmly entertained before having to worry about the strife of warfare. Apart from the capture of Hebb Greenwell on September 28 (the first recorded captured member of the Stuart Horse Artillery), the next greatest event to stir things up in the camps of Stuart's men occurred on October 1 when General Alfred Pleasonton's Federal cavalry crossed the Potomac near Shepherdstown. Whether or not the batteries of the 1st and 2nd Stuart Horse Artillery accompanied the repulse of Pleasonton is uncertain, though likely.

On October 8, General Lee issued orders to Stuart to "gain all information of the position, force, and probable intention of the enemy." He was to destroy the railroad bridge at Scotland, just a few miles north of Chambersburg, and was also authorized to destroy public property, seize horses and supplies as needed, and, in response to one of General John Popes repugnant orders of a few months earlier, to take hostage Federal, state, and local officials. Stuart issued orders for 1,800 specially selected men with the best mounts to rendezvous at Darkesville on October 9. The next great adventure was about to begin. In command of the elements of the expedition were Colonel W. H. F. "Rooney" Lee, Colonel William E. "Grumble" Jones, and Brigadier General Wade Hampton. Four guns of the 1st Stuart Horse Artillery under Major John Pelham would also accompany the expedition. The memoirs of a member of Hart's battery recalled that a section from his battery was also on the expedition. Leaving Darkesville at noon, the large body of troopers reached Hedgesville that night, where the contingent bivouacked.

At dawn on October 10, the advance across the Potomac was initiated by driving in the Federal picket at McCoy's Ferry. Crossing the river and reaching the old National Turnpike, the Confederate cavalry soon learned that it had just missed a force of Ohio troops under General Jacob D. Cox that had recently passed from Clear Spring en route to Kanawha via Cumberland. Moving on, the party was broken up as 600 men moved out along either side of the road to gather up all available horses from the area. By the time Stuart reached Mercersburg, the men were set to gather boots and shoes from the stores. Ironically, it was not until the merchants had received receipts for payment that they realized that the party was actually Southern. Gaining a good map of Franklin County, the party pressed on to the fields north of town, filling their haversacks with corn stripped directly from the fields. Moving now in a steady drizzle, Stuart finally reached Chambersburg by 8 p.m. Sending out scouting parties, Stuart placed the Stuart Horse Artillery a mile west of the town and demanded the surrender of the city. The intent of the Confederates was made clear to the mayor; all United States property would be used or destroyed at the pleasure of the general commanding. When asked who the general was, the aid refused to answer the mayor.

As the night pressed on, Stuart become greatly concerned about the torrential downpour wondering if he could recross the Potomac. Furthermore, the bridge they had been sent to destroy was found to have been constructed of iron. Early on October 11, the men were in their saddles and moved into the town's depots capturing overcoats, socks, underwear, and more than 5,000 rifles, pistols, small arms and sabers. The men burned the depots after gathering all they could carry. Stuart then ordered his group onward toward the little crossroads town of Gettysburg. Anticipating that Cox's Federals might be waiting for him at the point at which he had crossed the Potomac, Stuart decided on a longer route of return to Virginia. Driving his men even harder, the general pressed his force to Cashtown and turned sharply southward into Maryland, where the rules of property seizure were made null due to the Confederacy's attempt to continue to hold that state's trust. By the time Emmittsburg was reached, it was near dusk. Greeted by friendly citizens, the party paused only long enough to feed the horses. Stuart learned about a group of companies from Rush's Lancers, 6th Pennsylvania Cavalry that had recently passed through the town. With the capture of a Federal courier just a few miles south of Emmittsburg at Rocky Ridge it was also ascertained that the enemy was at least 800 strong under Pleasonton at Frederick. This force of enemy cavalry was at the moment en route to Mechanicstown (now Thurmont), only four miles west of Rocky Ridge.

With little time to waste, Stuart pressed his men on into the night with another 40 miles between them and the Potomac. As the column stretched out five miles long, snores of exhausted men could be heard among the clamor of sabers and accouterments in the dead of night. In front rode the 1st Virginia, followed close behind by Major Pelham and two guns of the Stuart Horse Artillery under Captain Breathed. Passing through Woodsboro, Libertytown, New London, and New Market, the column learned more of their pursuers and that General George Stoneman guarded the fords ahead at Poolesville with 4,000 - 5,000 men.

By dawn on the 12th the raiders had reached Hyattstown, just 12 miles above the Potomac. Again, to prevent an anticipated move, Stuart decided against a crossing at Noland's Ford at the mouth of the Monocacy River, or at Edward's Ferry. Instead, he would move to the rarely used White's Ford. Continuing forward, the column was sighted near Barnesville by Federal observers at a signal station on Sugar Loaf Mountain. To prevent further detection, Stuart followed the advice of Poolesville native Captain B. S. White in turning onto an abandoned road to the right of the main road. It would eventually take them to the main road that lead from Beallsville to the mouth of the Monocacy. With only a half mile more to go from that point, a farm road that would take them to White's Ford.

By 8 a.m., however, Pleasonton arrived at the aqueduct at the mouth of the Monocacy. Greeted with the sounds of picket fire in the direction of the Monocacy-Beallsville road, Pleasonton deployed two companies of cavalry and four of infantry. By the time this party had reached the Little Monocacy, it came within 200 yards of Stuart personally emerging from the old road onto the main highway. Whether the party hesitated due to some of the Confederates wearing blue overcoats or not, Stuart waited little before ordering a charge that sent the elements of the 8th Illinois Cavalry reeling to the protection of the infantry support companies from the 3rd and 4th Maine.

Map, Crossing of White's Ford, October 12, 1862

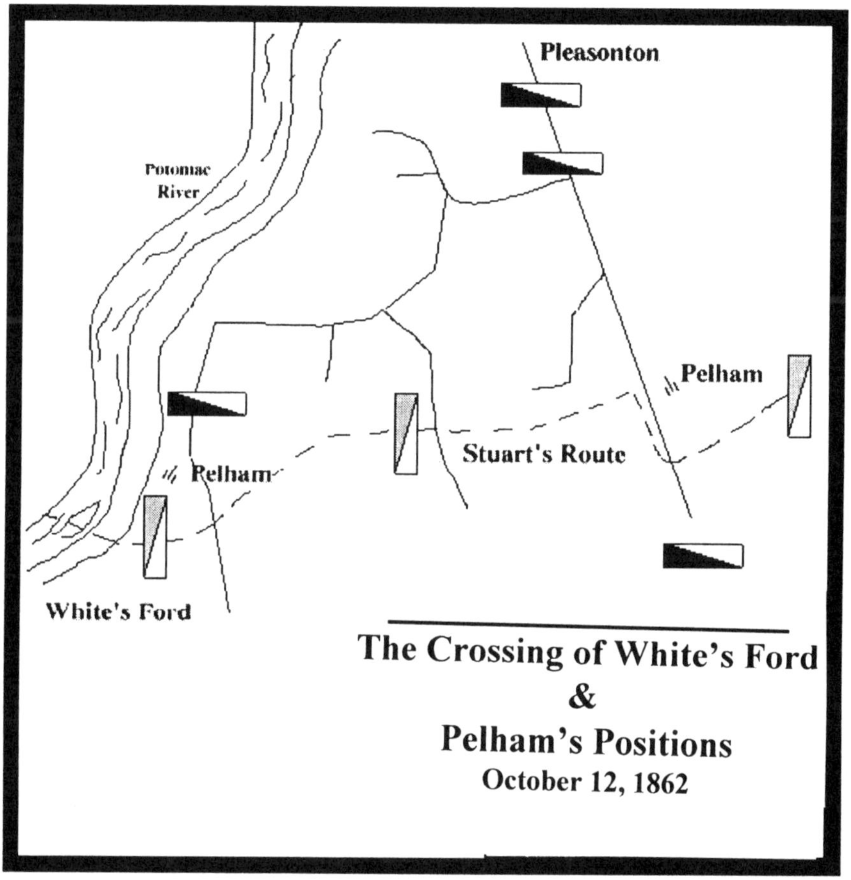

To aid in brushing off the enemy force, Pelham was called up with Breathed's to a ridge that commanded the road and the approaches to the escape route at White's Ford. The guns of the horse artillery and the fire of the dismounted troopers held open the way to Virginia.

In an effort to reconnoiter the approach, Rooney Lee sent scouts up to the canal and ford to ensure that the way would be cleared. Rooney's reconnaissance had been wise, for waiting at the point were some 200 men from the 99th Pennsylvania under Lieutenant Colonel Edwin R. Biles. With pickets posted along the canal and in and above the stone quarry overlooking the ford, the way was well covered by the enemy. Nevertheless, when he reported to Stuart, Rooney was ordered to force a crossing. Lee sent a flag of truce forward warning the enemy commander that he faced Stuart's entire cavalry force and that resistance would prove futile. With no response from the colonel, Rooney took other action by dividing up his men. However, when the ford was reached, Biles was found to have abandoned the position. The way home to Virginia was clear.

The columns surged forward, preceded by one gun of the Stuart Horse Artillery. By 9:30 a.m., Rooney Lee's columns began splashing across the ford. With the column extending for miles, Stuart began to be concerned with the rearguard under Colonel M. C. Butler. In order to keep the way open as long as possible, Pelham and some sharpshooters were ordered to withdraw in successive stages. With one gun already at the ford, one with Hampton, and another with the rearguard, Pelham's lone remaining piece was unlimbered on even higher ground a thousand yards to the west. From his high vantage point, Pelham was able to pour a devastating enfilade fire upon Pleasonton's skirmishers moving from the woods and up the ridge upon the road. Vital minutes were gained by the lonely gun. Expressing concern to Blackford, Stuart continued to worry about Butler's men. Blackford moved out and beyond the point where the couriers themselves feared to go and finally encountered Butler and urged him forward. Rushing forward with sabers drawn, the column was ready to cut its way through if necessary. There was also concern about losing the rearguard gun due to exhaustion of the horses. With the constant booming of Pelham's gun, the column was encouraged to move hastily onward, soon seeing the gallant major holding his position on the towpath of the canal. With the passing of Butler's column, Pelham closed the adventure as he brought his gun across to Virginia soil at last.

Exhausted physically and emotionally, Stuart's brave contingent took refuge in Leesburg and bivouacked for the night. In the days that followed, Stuart would head west into the mountains without one pursuer. As a celebration of their success in Pennsylvania and Maryland, another ball was held on October 15 at the "Bower."

The following day at dawn however, Union cavalry felt bold enough to come across the Potomac and drive in the Confederate pickets. Pelham and the Stuart Horse Artillery, along with Stuart, were soon on the road passing the "Bower" to Kearneysville. Bringing one piece onto a hill commanding the ground below, the gun opened on a brigade advancing on Lee Town. The ensuing encounter led to Stuart's withdrawal from position to position. Pelham, covering the retreat, annoyed the Federal advance under Brigadier General Andrew A. Humphreys. Finally, having enough, the enemy turned back. Against the possibility that the enemy might return, two guns of the Stuart Horse Artillery were placed on an elevated position for the night while men prepared rations. That same day, General Winfield Scott Hancock had also been advancing toward Charles Town. Though the fighting renewed briefly the next day, Stuart had been reinforced with infantry and artillery, squelching the Federal effort.

On October 26, McClellan finally started across the Potomac in force. Though east of the Blue Ridge, it was time for Stuart's troopers to once again earn their pay. Sending most of the mounted forces of the army to delay McClellan, Lee was prompted to move his main army toward Richmond again.

Early on October 28, Stuart's cavalry began crossing the Blue Ridge. Moving by way of Ashby's Gap to Millwood, the long gray column then continued toward a rendezvous near Bloomfield, just northeast of Upperville. Moving into bivouac at Seaton's Hill, Pelham and Lieutenant Walker spent the night by invitation in John Peyton Dulany's home at "Welbourne." Shreve remembered the time near Upperville as "recuperating in the midst of a rich district."

The cavalry was in action on October 31, and the days that followed would produce a string of small encounters with the enemy. Suffering from all sorts of problems with the horses, the cavalry would begin to feel hard-pressed this late fall.

On November 1, Stuart deployed the Stuart Horse Artillery with several regiments to Union Heights in order to hold off the Federal advance on Upperville and Paris. As the enemy cavalry began to probe with just a few men, Stuart felt comfortable to advance to Philomont. Skirmishing with Federal cavalry, Stuart again fell back to Union Heights and bivouacked that night. The fight was renewed on November 2. Breathed's 1st Stuart Horse Artillery had the most difficult time of it that day. Heavily engaged against Federal artillery, one of the battery's caissons was hit and exploded. Both C. D. Costigan and John Isaac Phillips were mortally hit. Costigan died immediately, while Phillips survived for another day before dying of the wounds he had received. Phillips was buried in Sharon Cemetery in Middleburg.

During the fight, Pelham personally went to one of the guns and took aim at the advancing color sergeant of the 7th Indiana Infantry. The shot found its mark at approximately 800 yards, killing the sergeant and the color corporal as well as wounding others. As the day progressed, the enemy advanced even more determinedly, forcing Stuart's withdrawal to Salem that night.

Following the affair at Union, the batteries of the Stuart Horse Artillery moved out along the road to Culpeper Court House via Sperryville. Somewhere along the way on November 3, 200 rounds of 3-inch ammunition and 500 friction primers were requisitioned and gathered by the Stuart Horse Artillery.

On November 4, while with Rosser's 5th Virginia Cavalry, Captain Henry's 2nd Stuart Horse Battery was nearly flanked at Barbee's Cross Roads by Averell's advance. Pelham adjusted to the situation and sent the guns to higher ground where they unlimbered again. The result pushed the enemy back temporarily; however, their next assault came through the woods from the rear. Totally taken off guard, Pelham quickly yelled out "Action rear!" and brought the guns about to face the onslaught from the rear with canister loaded in every shot. Some gunners mounted and rode into the fray, personally attacking their assailants. Meanwhile, back at the guns, the famous Napoleon detachment began to ring in with the "Marseillaise" and the music from the guns themselves. Withdrawing from the horrible fight of the day, Stuart's men camped that night south of the Rappahannock.

Sometime during the fighting of the "Ten Days," Lieutenant McGregor was seriously wounded at Goose Creek near Upperville and sent home to Alabama to recuperate. Lieutenant Ford assumed command in McGregor's absence. Shreve recollected that at Goose Creek, near Upperville, Lieutenant McGregor was "badly" wounded. "The writer with a man named Cook, helped to convey him to the ambulance, where we left him in charge of a surgeon, and attempting to find our way back to our guns we ran into the Yankees unwittingly, who fired a volley at us as we approached, over the hill, but we galloped out of reach of them, and rejoined our guns in a different place from where we left them."

Shreve recalled the fighting of the "Ten Days" as a deadly game of acquaintances:

> During the continuous daily fighting, the opposing forces became well acquainted with the make up and personality of each other. At night, around the camp fire, we discussed various conspicuous officers, apposed to us, as the "one of

the white horse," or "on the black horse," etc, and looked for and identified the same persons on the following days.

On November 6, across the Rappahannock at Waterloo Bridge, Pleasonton again came on, but was held in check by Pelham's guns and the cavalry. Rain and snow became the next opposition as the Confederate force withdrew and fought again at Gaines' Cross Roads and Newby's Cross Roads.

Finally, on November 10, Stuart set out to punish the Federal cavalry for the nine days of torturous fighting. What was planned to be a two-pronged attack on the enemy picket lines unraveled when the orders did not reach Wade Hampton. Therefore, with one of the batteries of the Stuart Horse Artillery and Fitz Lee's Brigade, under Colonel Rosser, Stuart struck the enemy at Corbin's Cross Roads, driving them back to Amissville initially. However, as the day progressed, things grew worse for the 1st Virginia in particular. As darkness approached, Rosser determined to retire.

For their service in the ten-day affair, Stuart had high praise for the Stuart Horse Artillery and his young major:

> The Stuart Horse Artillery comes in for a full share of this praise, and its gallant commander (Major John Pelham) exhibited a skill and courage which I have never seen surpassed. On this occasion I was more than ever struck with that extraordinary coolness and mastery of the situation which more eminently characterized this youthful officer than any other artillerist who has attracted my attention. His *coup d'oeil* was accurate and comprehensive, his choice of ground made with the eye of military genius, and his dispositions always such in retiring as to render it impossible for the enemy to press us without being severely punished for his temerity. His guns only retired from one position to assume another, and open upon the enemy with a fire so destructive that it threw their ranks into confusion and arrested their further progress.

Stuart reorganized his cavalry on November 10 into four brigades, but the horse artillery was not effected.

Using his "eyes" to figure out the enemy disposition, General Lee once again ordered Stuart to lead a reconnaissance on the Rappahannock toward Warrenton in the middle of the month. On November 18, Stuart advanced with Lee's and Hampton's brigades and the Stuart Horse Artillery. Encountering the enemy at Jeffersonton and shelling them at Warrenton Springs on the 19th, the reconnaissance force retired to Oak Shade that night and went into bivouac. Stuart had determined that the new Federal commander, General Ambrose E. Burnside, was concentrating his force on Fredericksburg.

For the remainder of the month, Pelham searched for more horses and harassed enemy gunboats near Port Royal. The horse diseases since the raid to Chambersburg were playing havoc among all units in the cavalry. Reorganization also took place prior to December, with Captain Henry's 2nd Stuart Horse Artillery assignment as a "headquarters section," and Captain Breathed's assignment to Fitz Lee's Brigade.

For a little while and to aid in the shelling of enemy gunboats, Captain John Milledge's Georgia battery was temporarily assigned with Pelham's headquarters (Henry Battery) on November 28. For ten days, the batteries would move along the river harassing and scouting gunboats. Robert A. Hardaway's Whitworth was also brought into action at Port Royal near the Pratt House with the 2nd Stuart Horse Artillery on December 5. A section of Moorman's Lynchburg Battery, recently redesignated as horse artillery, was also assigned to help with its two Blakelys. Rooney Lee reported that Pelham "waited until they [gunboats] were within 300 yards and opened with deadly effect, putting shot through and through one of the boats." The gunboats returned fire in the form of grape and canister which tore up the ground around the Lynchburg men and tore a leg away from one of them.

The shuffling around Fredericksburg turned serious by late on December 12. Coming into battery near General J. J. Archer's Tennessee Brigade on "Stonewall's" extreme right and almost at right angles to the Tennesseans, the Napoleon section of the 2nd Stuart Horse Artillery was to gain much credit during the following day. With Pelham wearing the red and blue striped ribbon given to him by observer Captain Lewis Guy Phillips of Her Majesty's Grenadier Guards Regiment, the young major was destined to receive high plaudits once again for his daring handling of the artillery. Shreve, assigned to the Napoleon section of the 2nd Stuart Horse Artillery, remembered of the following morning:

> The 13th of December, opened with a dense fog enveloping the whole field. A part of our Battery crept slowly from Hamilton's Crossing down into the plain, toward the river, until we came to a cross hedge row, of cedar, behind which we noiselessly formed "In Battery"... Beyond the hedge, we could hear the Federal Infantry (Franklin's Corps) maneovering; distinguishing a medley of voices, but could not see them. Evidently they were only a few hundred yards distant. The fog commenced to lift between nine and ten o'clock, and exposed to our view as we peered through the

> hedge, a grand spectacle of marshalled soldiery, in readiness for the fray, spread out in vast propositions, on the level plain, in our immediate front. With alertness, and yet fearing annihilation at their hands, in such close range of their infantry, we commenced firing. The unexpected presence of our guns so close to them seemed to paralyze them, and throw them into disorder. Instead of rushing for us, and overwhelming us with their numbers, they were evidently afraid of us, judging no doubt, that we had a strong force concealed behind the hedge.
>
> Immediately after our first fire, and while in the act of loading the second charge, we received the fire of their artillery, showing how ready their guns were for action. No. 1 in my detachment that morning, was "Hammond," who joined us in Maryland, during our recent campaign there, as he stepped in with the sponge staff to swab out, after the first fire, a shell from the enemy's gun cut him down, and he had time only to say "tell mother I die bravely." We continued firing for about twenty minutes, when we were commanded to cease, and to lie flat on the ground. While in this position, a shot struck squarely the head of one of our men, and decapitated him, as he lay prone to the ground. The rain of shot and shell upon us was terrific, both from their Field Batteries at close range, and also from their big guns on the north bank of the river, but being concealed by the hedge, the enemy's gunners shot too high and so we escaped. We resumed firing after a short rest, but were soon ordered to "limber to the rear," and we fell back, under cover of our guns, which opened from the hill in our rear, and left, as the enemy began to advance.

Though primarily aimed against enemy infantry, one shot from Henry's guns found its mark in an enemy caisson, sending it up in smoke. While the Napoleon kept an active fire against the enemy, Stuart felt that Pelham needed yet another gun to aid in his endeavor. John Esten Cooke was then ordered by Stuart to seek out a gun, and finding the Blakely from Henry's battery, ordered it out to Pelham's assistance. After going into battery and firing only one shot, the Federal artillery quickly found the range and took the Blakely out of action, leaving only the Napoleon to play upon the enemy lines.

With only a small detachment of cavalry to protect the Napoleon from an onrush of sharpshooters, Henry's one gun crew continued to deliver a deadly fire. In turn, the gun bore the brunt of between twelve to twenty enemy cannon. David Cardwell later remembered that "above the roar of their guns and the armament on Stafford Heights could be heard the voices of the French detachment singing the 'Marseillaise.'" Though proudly observing his young chief-of-artillery at the peak of his career, Stuart called upon twice to retire. "Tell the General I can hold my ground" was the young major's reply. Shifting his position from time to time, Pelham's gunners were taking severe losses. Pelham himself served the Napoleon near the close of the action. Finally, on the third message, and only after emptying his limber chest, Pelham drew his guns off the field.

Having observed Pelham's Napoleon in action, General Robert E. Lee was remembered as asking to what battery the gun belonged. When told it was Pelham's, his reply was, according to the *London Times*, "It is glorious to see such courage in one so young!" Likewise, in the same newspaper, Jackson was cited as stating "With a Pelham on either flank I could vanquish the world." While the quote may have been somewhat over dramatic, Cooke did overhear "Stonewall" asking Stuart "If you have another Pelham I wish you would give him to me."

Interestingly, while Pelham drew all of the praise, the young major later, according to John Cheves Haskell, "freely gave credit to Henry." Furthermore, rather than Pelham's refusal to retire, it was attributed by the major as Henry's decision. Having only lost two men by the first request to retire, it was Henry who justified staying in position on the flank.

Continuing after Pelham's withdrawal from the flank, Lieutenant Colonel R. Lindsay Walker's massed batteries on Prospect Hill roared down into the throat of the oncoming Union attack. Pelham himself was soon put in charge of several guns, possibly exceeding thirty guns, on the Confederate right flank to the right of Walker. While it is speculated that the balance of the Stuart Horse Artillery battalion was a part of this force, accounts are lacking in giving an accurate representation. Again and again the waves came on and were thrown back. One member of the Rockbridge Artillery, then stationed under Pelham, later wrote that Pelham seemed perfectly content in the fight "like a boy playing ball." Pelham's daring continued to play a factor until the close of the day. By 4 p.m. the horror had ended.

Considering a counterattack, like at Antietam, Jackson sent Stuart forward with cavalry and the Stuart Horse Artillery. Pelham unlimbered, fired, and limbered to move to yet another position to the front. The Federal counterfire increased. General Lee soon intervened and called off the counterattack. In preparation for the following day, Pelham, following Jackson's orders, dug in to protect the flank on the hill beside Hamilton's Crossing. With no onslaught on the following day, Henry's losses were limited to only six wounded and one killed at Fredericksburg.

Map, Fredericksburg, December 13, 1862

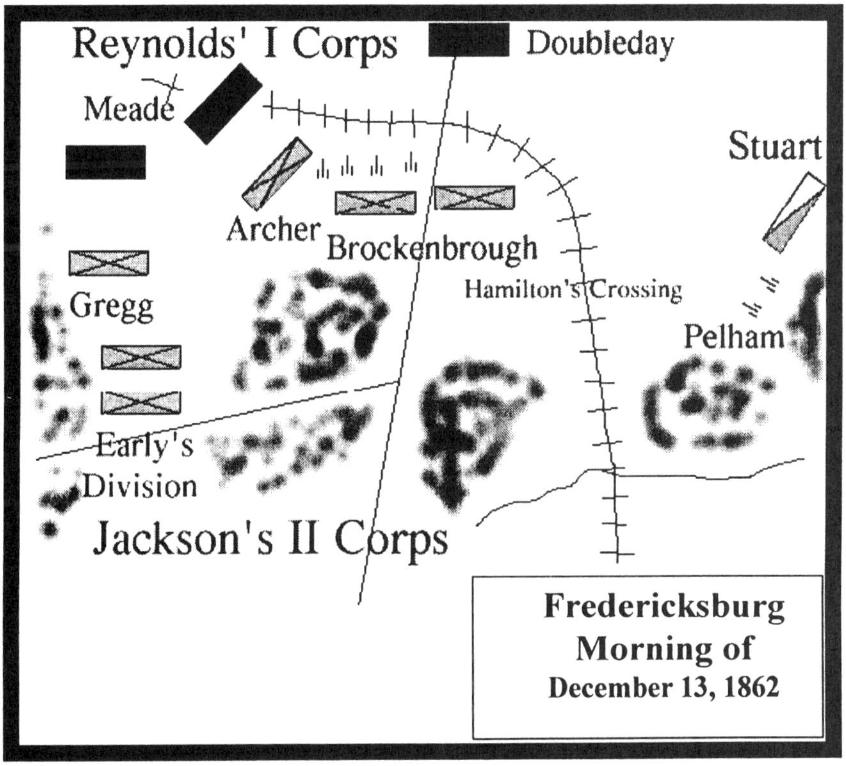

While Shreve remembered "Hammond" as being killed, the records show that David R. Barton was the sole victim that day. Among the men wounded were Samuel T. Evans and Joseph Henry Phlegar. Upon being received in the hospital, Evans, age 15, was seen as having taken a "Bomb-shell contusion of all internal organs over stomach. Greatly prostrated on admittance" After his recovery, Evans continued to serve, despite his age, as courier to General Robert E. Lee. Phlegar's serious wound by an enemy shell resulted in the loss of his right forearm. Ultimately taken to a converted tobacco warehouse in Lynchburg to recuperate, Phlegar would spend the rest of the war at home.

Moving into winter quarters must have been welcomed, but hard to believe, for the men of the 1st and 2nd Stuart Horse Artillery. Winter quarters were soon constructed at Camp-No-Camp near Tappahannock. Shreve recalled that "we built huts and also shelters for our horses, and settled down to enjoy a winter of repose." But the horse artillery was destined for more action before the close of the year.

On Christmas day, Pelham was called up with the Stuart Horse Artillery to move with the cavalry and reconnoiter the Federal lines. Stuart called upon Hampton to check the upper Rappahannock, while Fitz Lee and Rooney Lee were to select 600 men each. Pelham was to choose four of his best guns and rendezvous at Kelly's Ford the next day. Whether Pelham chose two separate sections from the 1st and 2nd Stuart Horse Artillery, or just one of the batteries, is left uncertain. Crossing the river at the ford and riding eight to ten miles before camping, the four select guns were in a less pleasant camp near Morrisville on the night of December 26. Directing Hampton to Occoquan, Fitz Lee to the Telegraph road between Dumfries and Aquia, and Rooney Lee between the two advancing along the right bank of the Quantico upon Dumfries, Stuart's objective was to take the Telegraph Road and capture all trains that might be passing.

At one point, while moving with Fitz Lee's detachments, Pelham beat the odds at a ford which everyone had previously believed impassible by vehicles. Only two and a half miles from Occoquan, Selectman's Ford was found a personal challenge for Pelham and was easily overcome.

When Rooney Lee reached Wheat's Mill, "where Telegraph road crosses Quantico Creek," a picket of about a dozen Federal infantry was encountered and charged by a squadron of the 9th Virginia Cavalry, which bagged the lot. This squadron, supported by two others, immediately pushed across, but were attacked by two full regiments of infantry posted at the edge of town, which forced them to retire. As they moved back, a squadron of the enemy's cavalry advanced, but were sent back upon being fired upon. Stuart wrote, "At this stage of affairs I reached the scene of action and ordered up [M. W.] Henry's battery of horse artillery, which opened with canister upon the enemy's position, and drove them from it." Enemy artillery soon came up to counter Henry, and a duel ensued. For the next several days the contingent made independent strikes into the enemy line, testing and feeling what Burnside was up to. By late on December 27 ammunition had been exhausted, which forced the guns to retire back across the Rappahannock and back to their winter quarters for the close of the year. But again, as Shreve recounted:

> ....in the midst of our contentment, we were hurriedly called to vacate our camp and proceed about six miles to the Rappahannock River, to intercept a gunboat ("the Line Yankee") that had gone up the river toward Fredericksburg. As she came down on her return, we opened fire on her, and we had a lively duel with her, and crippled her severely, but she escaped.

With a year of battles behind them, certainly the artillerists could put the war behind for just a few days or months before striking to the roads again. With the previous year having been so active, one would have to wonder what the following year held in store for the companies that "will never be quartered in one place long at a time."

Chapter 4

1863

"This service will be active and energetic...."

The chronicler of events in the 2nd Stuart Horse Artillery, George Shreve kept fond memories of his time in winter quarters near Tappahannock. But Shreve also recounted being forced again from their warm huts and into action against the enemy gunboats:

> When settled in winter quarters... in 1863, we were suddenly order[ed] to march to the river, to fight some gunboats. Dinner was just being cooked, and nearly ready, Rather than abandon the dinner, it was swallowed precipitately, notwithstanding many mouths and throats being scalded with hot soup. One man in attempting to swallow his rations, of meat, in one gulp, got choked, and in great agony, begged his chum to kill him to relieve his misery. He undertook to carry out the request with his pistol, but was restrained by others, until the surgeon succeeded in clearing his throat.... This illustrates the suddenness with which we often had to move, Often the kettle of soup was emptied, half cooked, and we would start off hungry and our hunger continued until satisfied from the "Hard tack" of the enemy.

George Shreve wrote of the messes that were formed that winter:

> There was nothing arbitrary in the composition of our messes. They constituted of from two to a dozen persons, according as they were drawn together, by accident or social preference. We had for a time a mess of one solitary individual, who on account of his filthy appearance or his mean disposition was abhorrent to his fellows, socially. On one occasion, when we were served with salt pork, ½ lb. to the man, the commissary, would ask "How many in your mess?" and deal out to the applicant the proper amount. This man applied, and in answer said "only one." The Sergt. Delivered to him ½ lb. Beginning at the end of the tail, and reaching to the place of union with the vertabrae. The look of disgust and disappointment displayed on his face as he went grumbling away was a picture not to be forgotten. The Sergeant told him to get into a mess if he wished to fare better....

> The messes went by different names, acquired by some incident, or general characteristic, and applied or dubbed by some wag, or christened by some friend.
>
> The "Foragers" were clever in pressing into their service whatever came within their reach, on the march, or by nocturnal raids, from our camp, upon near by farms. These even sold to other messes part of their booty, such as chickens, honey, fresh pork, etc. The purchaser enjoyed, and asked no questions for "conscience sake."
>
> The Lazy Mess was noted for the easy and careless-manner in which they moved along, with no apparent concern for the future. They seemed to let nothing disturb their enjoyment, trusting that their share would come to them.
>
> The "Industrious Mess," always alert, everything in its place, pans and utensils cleaned, and all in readiness for the next meal.
>
> The "Honest Mess" or the "Honesties" of which the writer was a member, so named because we expressed our disapproval of the manner of obtaining contributions, as practiced by some. Our mess had regular standing committees on particular duties. The Fire Committee to provide fuel for cooking, and warmth, a most important function. As soon as we would break ranks, after marching it may be until after dark on a cold day, and stopping where wood was scarce, the most active and forceful would secure the most wood. I have seen miles of worm fences demolished and put into piles by scrambling men, in five minutes. Each man would hold his pile down for his own mess. The water committee obtained....water and [must] have it in readiness as soon as possible. Sometimes in the darkness, he or they would have to go half a mile or more before finding it, and bring it to camp. The Bread Committee was expected to have the flour and salt etc., and the oven and pan ready, and as soon as the fire and water were at hand to bake the bread. The Meat Committee was to reserve and prepare the meat, so as to have it ready with the bread and to make coffee (of which we had very little) or substitute something else. The dishwashing Committee had to clean up after meals.

Pelham occupied himself during that winter reading a great deal, conversing frequently with others in Stuart's staff and caring for his three personal mounts. A great deal of time was spent by Pelham that winter with William W. Blackford, whose tent stood next to Pelham's. By the middle of March, the two would have made their way deep into Naiper's narrative of the Napoleonic Peninsular war.

During the period in winter quarters, the 2nd Horse Artillery lost their captain to promotion. Mathias Henry was destined for command of an artillery battalion in Longstreet's Corps. William McGregor was elected as a replacement in February. McGregor would lead the company until nearly the close of the war, when he too would be raised in rank.

Young "Willie" Hoxton also received a promotion during the winter. A brother of Pelham's West Point classmate, Llewellyn Griffith Hoxton, "Willie" had distinguished himself in action the previous summer and most especially at 2nd Manassas.

While Lieutenant Ford and a section of artillery accompanied General Rooney Lee in the attack and damaging of two enemy gunboats on the Rappahannock River on February 25, it wasn't until March that winter quarters were truly disrupted again. With a section each of Breathed's and McGregor's batteries, the 1st and 2nd Stuart Horse Artillery were represented with Fitz Lee's Brigade when he headed out to Kelly's Ford to repel General William W. Averell with his 3,000-man cavalry force. In the bustle of activity, Livingston Bennett was captured on March 4.

Pelham was not with the battalion when all the excitement had started. Just a few days before, Miss Nannie Price had sent Stuart some "candy stew" and reminded him to share some with "the gallant Pelham." Thinking it proper to thank the young lady in person, Pelham had ridden, with permission from Stuart, to Culpeper. Being no slacker of duty however, he would join the guns on the field on March 17 in time for the action.

In the early morning on March 17, pickets had opened with shots at Kelly's Ford, and by 8 a.m. Averell had crossed the ford. Fitz Lee reported: "I moved my command at once down the railroad, taking up a position to await their approach ordering my baggage wagons and disabled horses to the rear, toward Rapidan Station. Some time elapsing, and they not advancing, I determined to move upon them, and marched immediately for Kelly's."

As the Confederate cavalry arrived on the field, Fitz committed the detachments, each in turn making mounted charges across an open field toward the enemy position behind a high stone-and-rail fence. The Federals in response unleashed the firepower of their carbines and artillery. Sometime after the enemy guns had opened, Captain Breathed with his guns came up to the point where Fitz had fallen back a mile at the foot of a hill.

The ground was level to the front, approximately 600 yards across, bordered on the sides by woods. The road to Kelly's Ford ran directly up the middle.

Soon thereafter, the Federal artillery came up and unlimbered one piece in the road and two on the right, just a few yards apart. As the two sections of the respective Stuart Horse Artillery batteries blasted away, Pelham at some point rode up to Breathed and urged: "Captain, do not let your fire cease. Drive them from their position." In another instant Pelham was out with the 3rd Virginia Cavalry on another charge across Carter's Run. In that last furious ride of Major Pelham, he was struck from the saddle by a fragment. The piece of shell apparently struck the young major just wrong. Without damage to the youthful face, it had scored the rear of his skull and shattered the bone about two inches without penetrating, mortally wounding him.

As the day progressed, the Stuart Horse Artillery batteries, in true Pelham form, held their ground until Averell, exhausting his artillery ammunition, began to retire back across the ford. Though a service record shows him as wounded on March 7, it is most likely that James Mann was the only casualty of the 1st Stuart Horse Artillery that day. If the word of Pelham's wounding had not been received by the gunners before, it certainly must have reached them by now. No memoir, letter, or diary is currently known to exist that would give the initial shock of what Pelham's gunners must have experienced. Stuart, upon hearing prematurely that Pelham was dead, bowed his head on his horse's neck and wept.

After having been recovered immediately from the field, where he had fallen, Pelham's limp body was thrown across a horse as if already dead and taken back to Culpeper to the Shackelford home. The 1st Stuart Horse Artillery's guidon carrier, Robert Mackall, was mentioned in postwar years as helping Pelham off the field, though during that same time he was shown in a hospital in Richmond. Supposedly, owing to the rough handling of Pelham on his ride to Culpeper, the wound had been more aggravated. Sometime after midnight and near 1 a.m., the "Gallant" Pelham was dead. Major Heros von Borcke, a long-time fan of the cheerful blond haired boy, would accompany his body to Richmond.

The word of Pelham's death would not reach the town of Jacksonville, Alabama until April 2 when the *Jacksonville Republican* carried the horrible story:

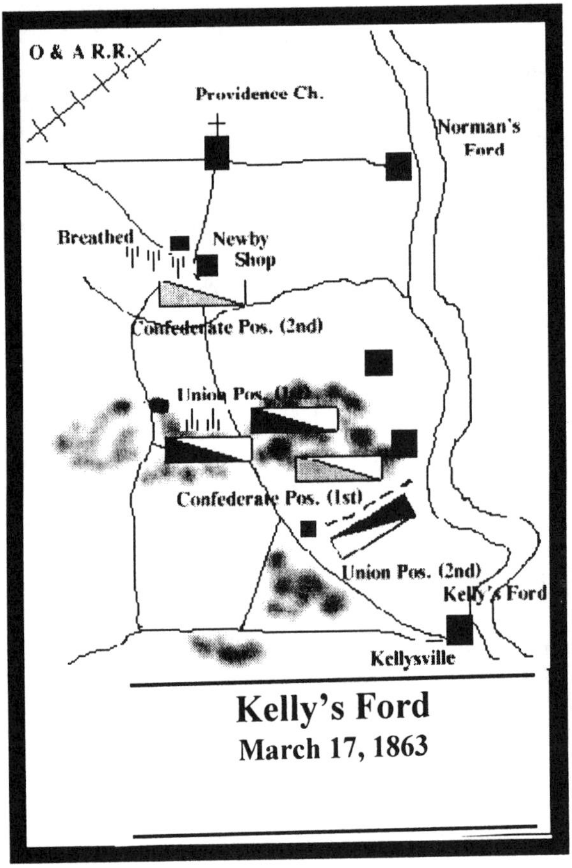
O & A R.R.
Providence Ch.
Norman's Ford
Breathed
Newby Shop
Confederate Pos. (2nd)
Union Pos. (1st)
Confederate Pos. (1st)
Union Pos. (2nd)
Kelly's Ford
Kellysville
Kelly's Ford
March 17, 1863

Major John Pelham

To show the estimation in which the lamented Maj. John Pelham was held in the army, we copy the following dispatch from Gen. Stuart, to Hon. J. L. M. Curry:

CULPEPPER, VA, 18th March, 1863.

To Hon. J. L. M. Curry:

The noble, the chivalric, the gallant PELHAM is no more. He was killed in action yesterday. His remains will be sent to you to-day. How much he was beloved, appreciated and admired let the tears of agony we have shed and the gloom of mourning throughout every command bear witness. His loss to the country, is irreparable.

J.E.B. Stuart,
Maj. Genl.

We also copy the following from the Richmond correspondent of the *Atlanta Confederacy*:

"The remains of the gallant Pelham were laid out in state in the Capitol, until they were removed under military escort, to the Petersburg depot, to be conveyed to his native State, Alabama. The body of the gallant warrior was visited by most of the ladies in Richmond, and the coffin was tastefully decorated with the choicest flowers, blended with evergreens culled by the estimable lady of Gov. Letcher, and placed on the coffin by her own hands."

Maj. John Pelham.

The sad gloom that has spread over our town, caused by the melancholy intelligence that the noble & gallant Maj. John Pelham had fallen, while bravely defending the righteous cause of our oppressed country, was somewhat alleviated by the announcement that his body would be consigned to its rest in our Town Grave Yard.

Back in the field in Virginia, on Friday, March 20, Stuart issued General Order No. 9 announcing the death of Pelham and proclaiming a 30-day mourning period. The men of the cavalry and horse artillery donned black armbands to honor the fallen Pelham. In one of his photographs taken shortly thereafter, Stuart was still wearing the black badge of mourning for his fallen horse artillerist.

John Esten Cooke remembered Pelham by stating:

> I never knew a human being of more stubborn nerve, or shrinking modesty. His blue eyes never fell before the stare of peril, but often when you spoke to him. His color never faded in the hottest hours of the most desperate fighting; but a word would often make him blush like a girl. A native of the great State of Alabama, he had the war blood of the South in his young veins; but I think he had come to love Virginia and the faces here with a love as warm as that of her own children. Virginia certainly loved him, her boy defender; but it was impossible to know him and not to love him. In that light blue eye was the soul of truth and chivalry. The smooth, boyish face was the veritable mirror of high breeding, delicacy, and honor.

Life following the loss of Pelham would be demanding, especially for his selected replacement. But when Robert Franklin Beckham was selected by Stuart as the replacement on April 8, 1863, the Stuart Horse Artillery Battalion gained an experienced and competent artillerist. A native of Culpeper, and prewar resident of Warrenton, Beckham, like Pelham, had attended West Point. Graduating 6th of 22 in 1859, he was initially assigned as a brevet 2nd lieutenant with the Topographical Engineers. On March 16, 1861, Beckham received a Confederate appointment as lieutenant of artillery. Serving initially with the Newtown Artillery, which later became the nucleus of the Stuart Horse Artillery, Beckham later served on the staff of General G. W. Smith. Like Pelham, he had ties, in a round about way, with Alabama. Assigned for service as drill master with the Jeff Davis (Alabama) Artillery, Beckham was elected as captain of the company, only to refuse the post. Again with General Smith as ordnance officer in August 1862, Beckham finally came to the Stuart Horse Artillery at Stuart's request. Apparently, however, it would not be until May that Beckham officially took charge as Marcellus Moorman held the post during the close of April.

With the Federal Army of the Potomac now under the command of General Joseph Hooker, activities along the Rappahannock began to come to life again in April. On April 28, Hooker crossed at Kelly's Ford. Concentrated at Brandy Station, Stuart kept Lee informed of Hooker's progress.

On the following day, Fitz Lee engaged the enemy cavalry at Stevensburg. Breathed's, McGregor's, and a section of Moorman's batteries were all active during the day. After firing a few rounds, the supporting cavalry charged and drove the enemy from the path before moving on to Raccoon Ford, where bivouac was made for the night.

The following day, prior to the opening of the battle of Chancellorsville, Lieutenant Wilmer Brown was posted with a gun of McGregor's Battery at Raccoon Ford. The enemy cavalry, under General George Stoneman, made an attempt to take the piece near Thompson's Cross Roads, crossing the river above the position and coming down from behind. Brown was successful in bringing off the piece, but in doing so lost eight men captured, one of whom was thought to be mortally wounded. The men being without horses and in an open field had no avenue of escape. Among those captured were George Freeman and William Tapp. However, owing to the lack of transportation, the prisoners were soon paroled and free from the threat of enemy prison camps.

With the bulk of the Stuart Horse Artillery later that morning, a grouping of six guns, including three of McGregor's, one of Breathed's, and two of Moorman's, were together near Catherine Furnace. Early in the evening, Stuart ordered Beckham to take four of the group to a position that Beckham reported was "indicated by Brigadier General A. R. Wright, with the view of driving back a line of the enemy's infantry from above the heights, about 1,200 yards in our front, that General Wright might occupy the ground with his brigade." Major Beckham brought the conglomerate of McGregor's and Breathed's pieces up to do the job. Beckham elaborated:

> It was supposed that the enemy had little or no artillery at this point, and that three or four guns would be sufficient for the purpose in view. The immediate effect of our fire was to scatter the enemy lines, and at the same time to draw upon us a storm of shot and shell from eight or ten pieces of artillery, well masked by the high, rolling ground on which they were placed. I caused the fire of our guns to be directed against these batteries, and, I think, with some effect, as it was not many minutes before the rapidity of the firing on the part of the enemy was so much diminished as to render it certain that some of his pieces had been compelled to retire... We continued to occupy the position for about three-quarters of an hour, when I received instructions from General Wright to withdraw the pieces.

The enemy fire was some of the worst the men of the batteries had ever experienced. McGregor's guns took the worst beating, having every man wounded but one at Lieutenant Burwell's piece; the axle of another gun was cut nearly in two. Shreve recalled that "the extra gun wheel carried on the caisson was brought up, and while my brother Richard, and one other man held up the end of the axle, two others put the wheel on, showing what strength may be developed under great excitement." McGregor and Lieutenant Ford were also present at the guns. Lieutenant F. H. Wigfall commanded the lone piece from Breathed's Battery. In all the loss was two men killed and six (including Lieutenant Burwell) wounded. McGregor also lost one horse killed and two disabled. Stuart's excellent staff officer, R. Channing Price, was mortally wounded by this same fire.

While Beckham kept busy assisting General Wright's efforts, Captain Breathed had been sent to accompany General Fitz Lee's Brigade on the left where two of his rifled guns engaged near Talley's farm, about a mile and a half above the furnace. Since the enemy force that opposed them was without artillery, Breathed's work was easy. According to locals, the loss inflicted upon the Federals was at least one hundred killed and wounded.

The following morning Beckham's battalion moved with Jackson's column to the Orange Turnpike, "within a few hundred yards of where the enemy's right flank rested." Beckham, by orders of General Stuart, pressed two sections forward in anticipation of General Stonewall Jackson's flank attack. Breathed led the advance section, under Johnston, forward, followed by a section of McGregor's Battery under Lieutenant Burwell. Shreve recalled that "strict silence was enjoined on this march. When a horse would attempt to neigh, he was struck on the nose to silence him." Breathed was immediately placed on the narrow turnpike and waited for the advance of Jackson's infantry. Only two pieces could be placed on the road. Shreve wrote: "The road was so narrow, and there was so little open country, that only a few guns were ordered to the front." Breathed's two other pieces were held immediately in the rear for reserve. For most of the opening action Major Beckham would ride close to Breathed's lead section, constantly monitoring the guns and the progress of battle. As the infantry line moved up, Beckham's guns continued the advance with the battle line, continuously stopping and firing along the route. The one difficulty that the guns met was fallen trees along the way cut by the enemy to serve as obstacles. H. H. Mathews remembered that "The roar of musketry was fearful. The smoke soon became dense, but above it all could be faintly heard and seen the rapid reports and flashing lights of Breathed's guns." Breathed constantly took the lead ahead of his guns and selected the best possible ground from which to deliver the deadly blows. Frequently, however, the enemy would find ground on which to place their batteries and deliver a "perfect hail-storm of canister."

Again Matthews recalled the activities of the late afternoon:

> The attack was fearfully terrific. The rout sudden and complete down the road went Breathed at a full gallop, men and horses and guns rushing pell mell over the first barricade and closing up on the backs of the fugitives, he gave the

command to unlimber, and into them with double shots of canister. There was havoc there - bones were not broken they were severed; flesh was not pierced, it was shredded; blood did not trickle, it left the victim in torrents.

At about 8 p.m. Beckham held up the batteries' progress owing to the darkness and great concentration of batteries opposing them about 1,200 yards distant on Fairview Heights near Chancellorsville. Matthews remembered that the "firing was kept up until a late hour in the night." Eventually, his cannoneers, exhausted and without rations for 48 hours, were retired to the rear. Meanwhile, a section of Moorman's Battery was brought up in the possibility that action might resume. By that time, however, Jackson's longer-range guns had come up and Moorman had no opportunity to participate.

The losses for May 2 included one killed each for McGregor's and Breathed's batteries. Breathed also lost two horses killed and five disabled. Postwar references show that a man named Corbin was the one of two killed from the 2nd Stuart Horse Artillery, while J. E. Duval was the other man. William Evans was reported in a postwar source as being killed from the 1st Stuart Horse. The only individual mentioned in the rosters as wounded was John Bini, on May 2. The native of New Orleans later had his leg amputated as a result.

On Sunday, May 3, the Stuart Horse Artillery battalion was held to the left of the Plank Road in case of an enemy flanking maneuver from the direction of Ely's Ford. The batteries remained in the same position until ordered to join General Fitz Lee to move with his brigade toward the ford. Here the batteries remained without occurrence until the march was taken to Orange Court House.

Sometime after the battle of Chancellorsville, the famous Napoleon gun of McGregor's Battery was buried. Shreve recollected:

> ....we buried a "Napoleon" Brass Gun, which we had kept in use through all our service, up to that time, and which did splendid work, on many a field; but it was always consider-ed too heavy for our flying batteries, especially as our hor-ses were not always the best. We received a 3 in. Gun to take its place. We buried it amongst the pines, near the road, and wrapped it in an old quilt as a shroud, which the writer had carried for a long time. So we bade farewell to our old "Napoleon" gun, and no man knows the situation of the grave.

The actions of the 1st and 2nd Stuart Horse Artillery, for the first time since their creation, are unclear from May 4 to May 20. It is likely that, as with Lieutenant Phelps' section of Moorman's Lynchburg Artillery, at least the 2nd Stuart Horse Artillery was in the area of Spotsylvania Court House on May 6. Arriving with Fitz Lee's Brigade, the artillery moved on the following day to Raccoon Ford and may have become engaged with the Lynchburg battery in its efforts to intercept enemy cavalry.

The 1st Stuart Horse Artillery recorded a number of men in the guard house in May. Exactly why some of the men were brought up on charges is uncertain. On May 11, Private John C. Phillips of the 2nd Stuart Horse began serving his sentence of performing the log drill for a day as a result of falling asleep on guard. Other sentences in the 1st Stuart Horse were those of James Mann and Thomas Burger. Mann also had to perform the log carrying duty. Burger, however, began his sentence on May 25 of "buck and gag." After being bucked for two hours at a time, he had to do another two hours of the log drill. While numerous others were also recorded in the guard house through June, no other sentences were clarified in the muster rolls.

On May 20, the batteries of the Stuart Horse Artillery battalion began to arrive at Brandy Station. On May 22, Stuart held a cavalry review, in which the 1st and 2nd Stuart Horse Artillery participated. For the last few weeks of May and into the first week of June, picket duty would become the routine for the horse batteries.

In the bright and beautiful morning of June 5, the Horse Artillery again joined in a grand review by General Stuart. After Stuart and a group of officers passed along the line twice, the squadrons of cavalry rode by the general and commenced a "charge" while Beckham's guns roared in unison. Little did they realize it, but in the firing of the guns, they confirmed Federal suspicions that Stuart was near Culpeper. Nevertheless, the review impressed Stuart so much that he insisted that another be scheduled for June 8 with General Robert E. Lee presiding.

Shreve recollected that there was the greatest respect among the cavalrymen for the horse artillerists:

> The Cavalry showed us great consideration, and seemed to proud of us. Frequently, when we would be passing to the front, after they had commenced the fight, they would cheer us, and exclaim "You boys will fix'em now," and they had great confidence in our ability to stem the tide and I think we never betrayed their confidence. We never had the enthusiasm after Gen. Stuart's death that animated us under his leadership.

With the confirmation of the cavalry presence that General Joe Hooker was hoping for, plans were immediately made for an attack. With nearly the bulk of the Federal cavalry force, Pleasonton moved in the direction of Brandy Station where Generals John Buford and David Gregg would make a two-pronged assault.

Following the review of June 8, the battalion, with its complement of four batteries, encamped in a wood on the Beverley Ford Road, near St. James Church and about a mile and a half from the river. Breathed's company was at the time detached and farther up the river with W. H. F. Lee's Brigade. With only Company A of the 6th Virginia Cavalry present along the ford as a picket, the men of the battalion bedded down in anticipation of moving with the cavalry the following day.

In the dark hours of the morning of June 9, the sleep of the cannoneers was broken when shots rang out at the ford; Buford was approaching. Shreve of the 2nd Stuart Horse recalled being one of the first to get up and rouse his sleeping comrades. Some heeded, while others dismissed the shots as of little concern. Meanwhile, Shreve and others attempted to gather the horses in the fields where they had been left to graze. Since Hart's South Carolina battery was camped on the extreme right of the battalion and nearest the ford, Beckham immediately ordered Captain Hart to place one gun by hand in the road to offer resistance while the other guns limbered up and moved back.

Soon the shots increased in frequency and the cavalry pickets came dashing through camp. Meanwhile, Hart's cannoneers, many still in no more than their shirt tails, blasted away with the single piece, causing the head of the enemy column to recoil temporarily. As the diligent Carolinians gave strong resistance, Lieutenant Colonel Thomas Marshall's 7th Virginia Cavalry and Major C. E. Flournoy's 6th Virginia Cavalry came up and allowed the rest of the battalion to move back six to eight hundred yards to the south, near the church, where they went into battery and commenced firing. General "Grumble" Jones at once saw to the continued protection of the horse artillery by calling up the full strength of his brigade to their support. Several of the battalion's horses were wounded before pulling out of the camp.

The position of the horse artillery was along the high ground stretching eastward from St. James Church and across the Beverley Ford Road. With such a grand line of sight the batteries had control of the road and a wide expanse to the north of the church. The first opposing force was that of the 3rd Indiana and part of Thomas Devin's Brigade. The approach of the Federals was immediately contested by the 11th and 12th Virginia and 35th Battalion Virginia Cavalry, ending in the Federal line being held up in the wood line.

Sometime after 8 a.m., Major Robert Morris' 6th Pennsylvania Cavalry (formerly Rush's Lancers) broke from the woods in the direction of the church. Beckham immediately opened at the range of approximately 800 yards, while another horse battery stationed near the Gee House tore apart the flank of the Keystone-state men. Bursting shells and flying bullets passed all around the Federal cavalrymen and they still came on. After a melee in the woods near the church, the Pennsylvanians were finally forced to retire. During the early morning action, three of Beckham's guns were disabled from the shock of recoils, leaving Beckham only five guns immediately with him. The others had been attached to Colonel M. C. Butler (one gun) on the Stevensburg Road, on the Kellysville Road (two guns), and on Fleetwood Hill (two guns under Lt. Carter of Chew's Battery) by order of Stuart. Three of the guns present remained with Jones' men and the other two pieces (McGregor's) were sent north of Brandy to assist in driving the enemy from that point.

Meanwhile, finding the position near the ford difficult to overcome, Buford shifted the bulk of his attack to the right and struck W. H. F. Lee's Brigade on Yew Ridge near the home of Dr. Daniel Green. At 8 a.m. Johnston's section of Breathed's Battery recrossed the Hazel River and took position on the road leading from Beverley toward Welford's Ford. Breathed's other section held near Starke's Ford. As the Federals came on they met fire from a section of guns under the command of Lieutenant Johnston and that of the dismounted brigade below the battery, posted behind a stone wall. After holding under the fire of four enemy guns, Johnston's section was finally ordered to retire at about 2 p.m. Joined by the other section, Breathed's guns took position again about two miles west of Brandy Station and continued a "desultory fire" until the enemy withdrew across the river.

For the rest of the battalion the action in the morning lulled for some time, but by about 10 a.m. a force of enemy appeared a short distance to the rear of what was left of the battalion along Fleetwood Hill and struck Lieutenant Bamberg's section of Hart's Battery head-on. At once the Confederates along with McGregor's two guns advanced upon the hill and quickly occupied it, while the enemy continued moving to the rear with Stuart's cavalry in pursuit. After the horse artillerists had placed the guns on the hill facing the river, the 1st New Jersey Cavalry appeared from the woods to the left and rear. The advance of the enemy was quickly passed by, paralleling the guns to the rear just beyond the unlimbered caissons. Friend and foe seemed to both pause for a brief time, realizing what had occurred without action.

Shreve wrote of the incident:

> We placed our guns on the hill, facing the river, and in a few minutes, a body of Federal cavalry emerged from the

> woods to our left, and rear, immediately upon us, and passed paralleling our rear, just behind our unlimbered caissons. The unexpected suddenness of this movement seemed to paralyze us all, friend and foe alike, for the enemy passed at a walk, accelerated to a trot, and did not design to fire upon or charge us, or attempt to make us surrender. We were, indeed, practically in their power, as the guns were pointed in the opposite direction. As an organized body, we failed to do anything, but each man acted according to his own impulse.

The battalion was essentially at the enemy's mercy. All at once fighting at close quarters broke out. Lieutenant Ford, being mounted, took a prisoner and with the foe's pistol shot one of the other cavalrymen. Lieutenant Hoxton killed another.

The details of Edward Sully's capture of a trooper were much more publicized, and dramatically portrayed, in the July 7, 1863 issue of the Staunton *Spectator*. One Federal horseman ordered Sully to surrender. When Sully refused, the trooper leveled his pistol at the private, only to find his rounds exhausted. Drawing his sword, the cavalryman "dashing his spurs into the horse, tried to cut Sully down; but our hero was ready for him...." As the trooper wielded his sword to strike Sully, the artillerist ducked, "and as the horse dashed past," taking hold of his sponge staff with both hands, " swinging it around his head, he brought it down with all force on the back of the Yankee's head, killing him instantly and tumbling him headlong from his horse, of which, with the accouterments, he took immediate possession. For his actions, the *Spectator* stated that Sully was later "highly praised by his immediate officers and by Gen. J. E. B. Stuart, who has mentioned him favorably to Gen. Lee, who speaks of him in the highest terms."

Shreve recalled that another man "with a trail hand spike captured another prisoner." Shreve himself "with no weapon, tried to capture one with words only, but he heeded not, and why he did not shoot me down as I stood on the ground looking up at him at a distance of a few feet only, and why [they] did not exterminate our little band of cannoneers, none of them being armed? Probably their commander realized that he was in a tight place, hemmed in by our cavalry, and he had not time to bother with us; for they had only ridden a short distance, from us when they were charged from another direction, by one of our regiments and routed." The passage through McGregor's and Hart's guns cost the New Jersey regiment the mortal wounding of their commander and his executive officer.

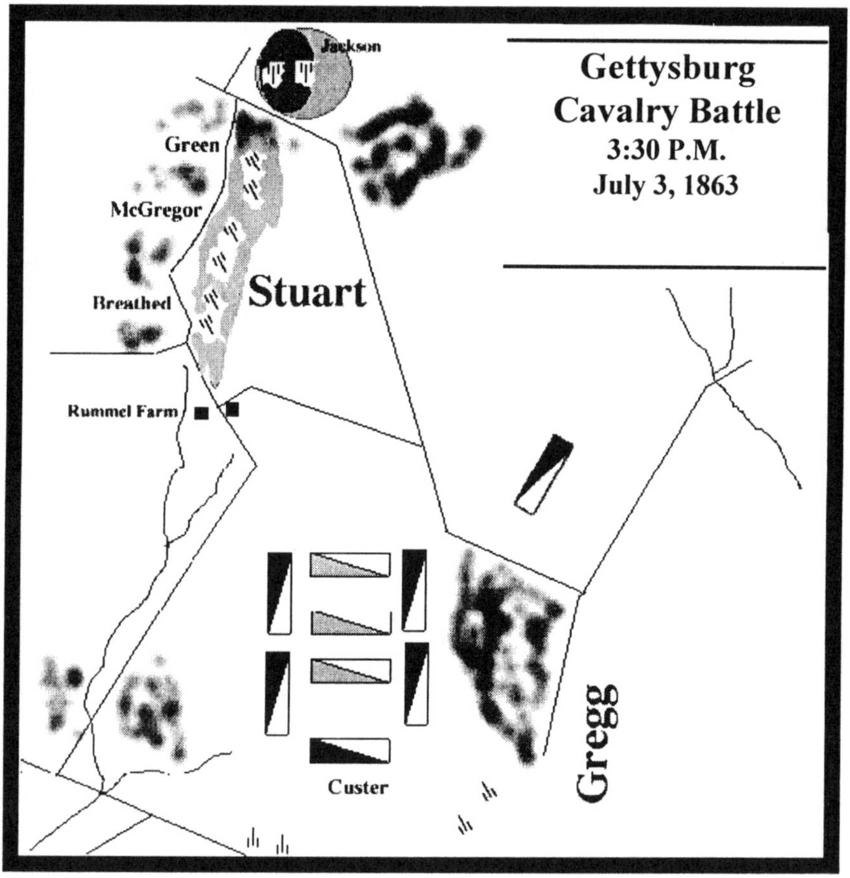
Gettysburg
Cavalry Battle
3:30 P.M.
July 3, 1863
Jackson
Green
McGregor
Breathed
Stuart
Rummel Farm
Custer
Gregg

The casualties for the battalion for the day were relatively light, though costly in the toll of brave men. Beckham reported: "wounded, 3 privates; missing 1. Wounded-1 of Hart's battery, 3 of Chew's battery, and 3 of Breathed's battery. Total-Killed, 1; wounded, 10; and missing, 1."

Postwar records cite Harry Wiches and Harry Wagner as two of the three wounded in the 1st Stuart Horse. An interesting side bar to the day was the fact that Samuel Lanier was to begin his furlough on June 9. If he ever had the opportunity or was able to enjoy it after the battle was unrecorded.

From June 13, Breathed's Battery accompanied by two guns of McGregor's battery moved with W. H. F. Lee's, Fitz Lee's (temporarily under Colonel Thomas T. Munford of the 2nd Virginia), and Hampton's brigades around the enemy. Providing a screen of pickets to protect and shield the Army of Northern Virginia in its advance northward, the cavalry continued on the following day through Amissville to Gaines' Cross Roads before camping. One man was also recorded on the same day from Breathed's Battery as being captured. The following day, Munford's Brigade marched through Markham, and then farther north to Somerset Mills, just south of Paris, where it bivouacked for the night. Stuart himself followed with the rest of the cavalry, continuing to screen the infantry that was moving on parallel roads to the west. On the night of June 16, Munford's men were bivouacking between Piedmont Station and Upperville.

On June 17, the long cavalry columns passed through Paris and Upperville. Stuart, continuing to probe the enemy to find their disposition, directed Munford to march on Aldie. Taking the 1st, 4th, and 5th Virginia and Breathed's Battery to Dover, three miles east of Middleburg, the contingent halted to water its horses in the village.

As had been the tradition for the past few months, McGregor's two-gun battery may have been accompanying Breathed, as manpower had been severely reduced in the 2nd Stuart Horse Artillery recently. Lieutenant Wigfall's letter of July 7, 1863, mentioned that: "Since the 13th of June inclusive there has not been a day on which we have not marched. Our battery and two guns of McGregor's were with the cavalry Fitz and W. H. F. Lee's brigades and Hampton's on the expedition round the enemy."

In the meantime, Rosser went ahead with the 5th Virginia to Aldie. By 4 p.m., Rosser was en route to Aldie Gap where he hoped to reunite with the rest of the brigade. However, Federal cavalry that had pushed Munford's advance detachment unhinged the plan. Advancing mounted charges against the Confederates, the enemy cavalry took severe losses. Positioned near the 1st Virginia just above the Little River Turnpike, just half a mile from the DiSierga farm, Breathed's Battery proved extremely deadly to the 2nd New York and 6th Ohio. Losses among the 1st Stuart Horse Artillery totaled two killed and three wounded. According to the rosters, Hebb Greenwell of the 1st Stuart Horse was killed on June 15 near Rector's Cross Roads.

For the next few days, fighting continued from Aldie to Snickersville, Mountsville, and on June 22, at Middleburg or White Pump once again. Shreve of the 2nd Stuart Horse Artillery described the fighting at the "Blacksmith's shop:"

> At this time we were retiring our guns from where we had been firing to another position to our rear. But for this successful change, we probably would have lost some of our guns. Taking a new position, we opened fire again, but as the enemy was in overwhelming force, we were forced to retire from one position to another along the road. At one position, as we were firing from the near incline of the hill with the guns pointing over the crest, (to secure more safety) the recoil sent one of the guns back down a steep hill for about 100 feet, and we worked like Trojans getting it back up the hill into place again, and entirely on top of the crest. Our detachments were reduced in the number of men composing them, thus rendering the work very laborious to the few. At Stone Bridge, near Rector's, we made a determined stand, and had a lively artillery duel from the hills opposite. At this place the writer was wounded in the leg, but continued with the Battery on the retreat through Upperville toward the Gap.

Now with the main portion of the army past him, Stuart was clear to march on into Maryland, with the ultimate goal to reunite with General Lee in Pennsylvania. The three brigades of Fitz Lee, W. H. F. Lee, and Hampton, along with six guns from the 1st and 2nd Stuart Horse Artillery, marched to Salem on the night of June 24. The following day, the journey continued through Thoroughfare Gap, where a scrap was made with General Winfield Scott Hancock's Federal infantry. Unable to pass through that point, Stuart moved instead through Glasscock's Gap and on to Gum Springs to bivouac for the night. That evening, the horse artillery bombarded the enemy wagon train for a while.

Avoiding the long columns of Federal infantry on their advance northward, Stuart opted to move his own brigades through Brentsville to near Wolf Shoals. From Somerset Mills, the cavalry continued toward Fairfax Station on June 27. Gathering prisoners and sutler stores, the cavalry pressed on through Dranesville and crossed the Potomac at Seneca Ford. Blowing up the locks of the Chesapeake and Ohio Canal, Stuart also captured a train and skirmished with a home guard unit on the trek. From Rockville, Stuart led his men on through Cooksville and finally Westminster.

Upon reaching Westminster, Lieutenant Halsey Wigfall of the 1st Stuart Horse Artillery remarked that he did not resemble a Confederate officer much. His boots were worn out and his pants and coat resembled a beggar's. The crown of Wigfall's hat was also gone.

Several men were captured beginning on June 30. In the 2nd Stuart Horse, Reuben Bugler was the only man, being exchanged before the close of the war. The 1st Stuart Horse, however, lost men in greater numbers. Martin Cahill, James McNelis, John Murphy, and Thomas Rudden were all recorded as captured on June 30 at Westminster.

On June 30, Colonel John R. Chambliss' Brigade led the advance toward Hanover, Pennsylvania. Operating on the left flank, Fitz Lee's Brigade moved between the Littlestown and Hanover roads. Before noon, the Confederates encountered a long column of Federal cavalry which turned toward Stuart's columns and drove Chambliss' men back. In a short time, Stuart found the numbers incredibly overwhelming. Among the assailants were General George Armstrong Custer's and Judson Kilpatrick's brigades.

After breaking off the engagement at the Forney Farm, there was a brief lull while Stuart began placing Chambliss' Brigade, along with guns from Hart's and McGregor's batteries, on a ridge extending from Keller Farm, southeast of Pennville to Mt. Olivet Cemetery. Meanwhile, Fitz Lee arrived with his brigade with the four more guns of the 1st Stuart Horse Artillery. Setting up near Littlestown on the Frederick road, the Confederates began to threaten the Federal right. In the artillery duel that followed, Beckham's batteries challenged the Federal horse artillery west of the town on Bunker Hill for more than two hours. The adversaries were Lieutenant Alexander C. M. Pennington's Battery M, 2nd U.S. Artillery, and Lieutenant Samuel S. Elder's Battery E, 4th U.S. Artillery. At 2 p.m., another lull occurred and the engagement was broken off. As Kilpatrick drew off, Stuart retired his cavalry toward Dover, Pennsylvania.

Riding throughout the night, the Confederate troopers reached the town by daybreak. The captured wagons had become a true burden. With mules that had not eaten or been watered in over three days, the column was dragging. Unable to find Lee and the main body of the army, Stuart went on a rumor that they were in Carlisle and rerouted his cavalry. Of the journey through Pennsylvania Lieutenant Wigfall wrote:

> You should have seen the Dutch people in York County turning out with water and milk and bread... and apple-butter for the "Rugged rebels".... In two or three instances I

> found people who seemed really glad to see us and at scores of houses they had refreshments at the door for the soldier. The people seemed generally not to know exactly what to expect and I don't think would have been at all astonished if every building had been set on fire by us... nor would a great many have been surprised if we had concluded the business by massacring the women and children....

As the columns trudged along in the hot summer sun, men found little more relief than on their trek to Dover. With Fitz Lee in the advance, the Confederates first encountered a large body of Pennsylvania militia at the edge of Carlisle. Stuart directed Lee to send a flag of truce requiring unconditional surrender or suffer an artillery bombardment. Major General William F. Smith refused and barricaded the main streets. His men held their own fire as the Stuart Horse Artillery lobbed more than 80 rounds into Carlisle. The affair lasted until midnight and resulted in the wounding of twelve militiamen, killing of one horse, and damage to a number of houses. Detachments had been ordered to burn Carlisle Barracks, which was accomplished along with the destruction of a few other buildings.

Stuart, finally making contact with Lee, was ordered to report immediately to Gettysburg. At 3 a.m., the Confederate horse artillery fired three parting shots toward Carlisle as the cavalry began to move on. Worn out and jaded, the Confederate cavalry pressed on through Papertown and Petersburg before reaching the main body of the army on the afternoon of July 2. For the duration of the day, both men and mounts were given an opportunity to rest. Lieutenant Wigfall recalled in a letter that during the time from Union to Gettysburg, the harness had been off the horses only twice.

By 6 a.m. on July 3, Stuart's cavalry was on the march once again with the goal of striking General George Gordon Meade's Army of the Potomac from behind. Riding out upon the York Road, Stuart advanced to the point behind a wooded ridge. The cavalry had gone ahead with the batteries of Thomas E. Jackson, William H. Griffin, and a section from C. A. Green's Louisiana Guard Artillery. The 1st and 2nd Stuart Horse Artillery remained behind, having a serious depletion of ammunition. It wouldn't be until afternoon that the balance of the Stuart Horse Artillery battalion would be on the field.

By 2 p.m., both Breathed and McGregor arrived in the middle of an action that had opened much earlier in the day. Breathed's 1st Stuart Horse Artillery brought with it three 3-inch rifles. McGregor's 2nd Stuart Horse Artillery, having been replenished on the journey to Pennsylvania, arrived with its two Napoleons and two 3-inch rifles. Positioned near the Rummel Farm, Pelham's old gunners were soon tangling with both Pennington's Battery and Captain Alanson M. Randol's Batteries E/G, 1st U.S. Artillery. The Confederate batteries held their own until ammunition had again been exhausted. The batteries were withdrawn before dusk. Losses in Breathed's Battery included six killed and eight wounded. Horses killed and disabled numbered ten. McGregor had nearly the same number of casualties with five killed, seven wounded, and eleven horses wounded or killed.

On July 4, Lee had Stuart moving along the roads to protect the main army's retreat. Through pouring rain, the long march back to Virginia was initiated. Fitz Lee's Brigade acted as the rearguard for the army and was the last of the Confederates to withdraw on the morning of July 5. The following day, Fitz Lee arrived at Williamsport, in time to save the army's wagon train from being captured. On July 7, Fitz concentrated his brigade at Funkstown.

Stuart, ever itching for a good fight, drew his cavalry together and struck the enemy near Boonsboro on the following day. As ever, the 1st and 2nd Stuart Horse Artillery, with Moorman's Lynchburg Battery, chimed in with their guns as well. After a good duel of cannons, the enemy batteries were driven from the hill by McGregor's Napoleons, "which, for close fighting, evinced this day their great superiority over rifled guns of greater number." All was not a glorious feat however, as losses in Breathed's Battery included one enlisted killed and four wounded. McGregor also came off the field with at least one wounded. Ashton Chichester was one of the men wounded. He reached the General Hospital in Charlottesville on July 9. Chichester returned to duty in just over two weeks.

On July 9, Stuart sent his troopers out to check the enemy once again, but finding them in strength, drew back to Funkstown. The following day the Federals felt the same boldness and probed Stuart, forcing him to fall back. In a letter dated July 12, Lieutenant Wigfall wrote his sister of the condition of Breathed's Battery and stated that the horses were worn out and the ammunition exhausted. "All of the guns are now in the same condition, but the Ordnance Officer of the Division, Capt. Jno. Esten Cooke, perhaps better known to you as Tristan Joyeuse, Gent., has sent to Staunton for ammunition." Wigfall also mentioned that, without a horse temporarily, he was "for the present with "Company Q." So few horses were available, in fact, that before recrossing the Potomac, one section had to cross alone and leave the one with fewer horses to protect the rear. By July 14, the daily clashes and artillery duels were ended temporarily as Stuart crossed the Potomac at Williamsport.

Before, the 1st Stuart Horse Artillery made its way from Maryland, Captain Breathed took a short opportunity to visit his father's house. Matthews remembered:

> ...he left his command and with his friend ventured to call at his old home. In passing the College grounds he fell under the observation of a distinguished prelate, whose sustenance and reputation had been generously furnished him for years by a Southern constituency. He had not yet dismounted from his horse, before he observed this distinguished disciple of "peace on earth and good will towards men" breaking towards ___ Road where some Federal Cavalry had been lurking during the day, and before many minutes had passed they were upon him, but he galloped definatly away much to the disappointment of his pursuers, and a little to the chagrin of the distinguished prelate, whose loyalty in this behalf and other similar and redoubtable manifestations of cheap patriotism, have since been generously rewarded with an Episcopal hat. Breathed often laughed heartily over his old preceptor's first effort in military strategy.

By the close of the campaign Breathed's company had recorded ten missing and McGregor, five. Of those men captured, the rosters identify a few. In the 1st Stuart Horse were mentioned Patrick Murphy (7/1), Sterling Murray (7/2), Charles Seymour and John Warro (7/2, Carlisle), Thomas Burger (7/2, Westminster), James I. Higgins (7/4), Julius Bates (7/5, Westminster), Stephen Mays (7/6 at Hanover), James F. Brown (7/7), Patrick Morgan (7/8-14), John Smith (7/8), Edward King and Henry Weeks (7/12, Hagerstown), and William Swancoat (7/13, being left behind sick). Of that number captured, most would later be exchanged by March 1865 and return to service. Julius Bates, however, opted to join the enemy by August 30. Another one of the captured, James F. Brown, had been a civilian courier for J. E. B. Stuart in 1861. He would return and would be present at Appomattox in April 1865.

In the 2nd Stuart Horse, far fewer were listed as captured, including John Lumpkin and John Durrant (7/4 at Gettysburg). Durrant later opted to join Ahl's Federal battery on July 27.

Undoubtedly, upon their return to Virginia the men must have felt safe again. Some men went out looking for some of the sweeter things that the men loved about the state they defended so often. Shreve wrote: "In one of our detachments, there was a man named Wingfield, who had served in the U. S. Regulars. He was a member of the mess, called the "Foragers." His forte was to bring into camp beehives, which he would take anywhere along the road, and travel with them on his horse for many miles, and he was never known to be stung."

No sooner had the Confederates crossed the Potomac than the Federal cavalry started in pursuit. On July 15, learning that the enemy had crossed the Potomac, Stuart devised an attack intended to drive them back for a while. After ordering Fitz Lee to concentrate between Charles Town and Leetown, Stuart was called away, leaving Fitz in overall charge of the affair. He met with success during the following day; however, at one point he was about to be driven from the field when Chambliss arrived in time to send the Federals retreating. No mention was made of horse artillery participation. Thomas Loudenslager, 1st Stuart Horse, was listed as wounded on July 15. Peter Vinne, 2nd Stuart Horse, may have been wounded also during this time, though shown in the rosters as wounded on July 22 at Funkstown, "blown from a muzzle of a gun." Another possible loss in the 1st Stuart Horse was Private N. Coit, who was listed as wounded and in the hospital on July 27.

While General Meade was determined to cross the Potomac again, Confederate commanders were not surprised when he moved his forces east of the Blue Ridge on July 21. Apparently having taken a different route than their respective brigades, the horse artillery was located farther to the middle of the Shenandoah Valley when Stuart began to move. On July 24, while portions of the horse artillery battalion including sections of McGregor and Moorman moved from their position in Page County toward Gaines' Cross Roads, the bulk of the battalion passed through Thoroughfare Gap. Shreve recalled of his section: "We were soon on our way up the Valley, crossing the Blue Ridge at Luray Gap."

Once in Culpeper County, the battalion set up camp on the Botts farm until July 31, when Breathed, with three guns from his battery and a section from the Moorman's Battery, was ordered to Brandy Station. When they arrived at Ellison's Ford on the Rapidan River that night, they found the river too high to cross without removing the ammunition from the caissons. With the assistance of the cavalry this was accomplished and the ford crossed. From there the contingent moved toward Fredericksburg, via Roane's Church, where time was allotted to feed and rest the horses. Finally arriving at Salem Church, the artillery contingent remained until August 3.

Shortly before dawn on August 1, Hart's Battery, along with at least a section of McGregor's Battery, attached to Hampton's Brigade (temporarily under Colonel Pierce M. B. Young), was up shortly after dawn. General John Buford's 1st Cavalry Division had crossed the Rappahannock and was heading for Brandy Station. Moving his brigade northeast along the railroad tracks in one of the hottest days of that summer, Young drew up his men perpendicular to the Orange and Alexandria tracks. The batteries of the horse artillery were placed on the left of the line, just north of the tracks. Young opened the battle with his horse artillery, but the Confederate gunners were hard pressed to keep up a steady return fire. When Buford commenced the attack, the Confederates were driven almost to Culpeper, and at one point Yankees began to close in around Young's flanks. In a desperate attempt to break out of the tight position, Young called upon the horse artillery, which opened with canister at less than fifty yards, allowing temporary relief and a chance at escape. The first section from the Ashby Battery was also involved in the action. Shreve recalled the day:

> ...a large body of Federal cavalry crossed the Rappahannock, and attacked us, and pressed us back slowly toward Culpeper Court House. The day was very hot. At a point about a mile south of Brandy Station, being pressed very closely, and in danger of being overwhelmed by numbers, both in front and on the left flank, we gave them canister at very short range. The order was given, "A double charge," and for the first time (according to my memory) we loaded a double charge, and let them have it. This repulsed them, and gave us time to "limber to the rear."
>
> On nearing Culpeper, some infantry [from Anderson's division] came to our help, and the tide of battle was reversed and the enemy driven across the river. We found on the field of conflict, after the foe had retreated, one of our men (Private Lacking), who had been badly wounded, whom the enemy had cared for. They had erected over him a temporary shelter, and had kindly administered to his wants, as best they could. Our Lieutenant Burwell was also wounded in this battle. The writer accompanied him to a private house in Culpeper, remained with him that night, and saw him aboard a train the next morning, bidding him "bon voyage." He seemed very hopeful of getting well, and returning to us soon, but alas; we never saw him more, for at his home, near Millwood, Clarke County, he died.

Losses in the 2nd Stuart Horse, other than Burwell, included Private Lacking and Robert B. Russell. When Robert Burwell was laid to rest in the Old Chapel Cemetery in Clarke County, no one guessed that his former battery commander, Mathias Henry, and ordnance officer, John Esten Cooke, would, after the war, also be laid to rest nearby.

The same day as the fight at Brandy Station, Lloyd Beall entered into the ranks of the 1st Stuart Horse. Beall, after fourteen years in the U.S. service, eleven of which was in the artillery, had decided to make his way south. The Marylander, in a letter in February 1864, wrote, "I frequently refused to do duty in the United States service, especially after the first battle of Manassas at which time I said that I would never again take up arms against the South for which I was placed under arrest and was not released until I consented to do the duty of mustering officer." Beall also wrote that because of his wife, it was impossible to travel, thus compelling him "to perform that duty." Eventually leaving his post as mustering officer in Kansas, he arrived in Virginia for service with a battery that would find his experience useful. Beginning as early as November, Beall would have endorsements written on his behalf for a commission from Stuart, Fitz Lee, and General W. C. Wickham.

Apart from the picket duties near Brandy Station (Ashby Artillery gunner George Neese recounted the sections of Chew's battery alternating outpost picket duty with sections of McGregor's 2nd Stuart Horse from August 20-22; while Shoemaker recalled the same for his battery and McGregor's toward the close of the month), for the duration of the month, there were fights all along the Central Virginia landscape. At Brandy Station, Kelly's Ford, and Culpeper. However, the flurry of activity leaves void any histories of the activity of the 1st and 2nd Stuart Horse Artillery. The 1st Stuart Horse had some losses recorded for the month. Anthony J. Diller was listed as wounded at Smithfield and Moses Febry was shown as captured.

Learning that Lee had detached Longstreet's Corps to Georgia, on September 13 Meade felt confident in striking into Confederate country, sending his cavalry toward Culpeper Court House. Confederate cavalry of Lunsford Lomax's Laurel and W. H. F. Lee's brigades and McGregor's, Chew's, and Moorman's artillery, clashed with them near Brandy Station.

Samuel McClearn, of the 2nd Stuart Horse, was captured at Culpeper during the activity on the 13th. The following day, John DeMaine and Robert Hunter would suffer the same fate, only to take the Oath of Allegiance by July 1864. DeMaine would, however, return to the ranks of the company by Petersburg in 1864. Yet another, George Field, fell into enemy hands on September 16, later to be exchanged.

Shreve wrote of the activities of the 2nd Stuart Horse Artillery in the affair:

> The enemy advanced slowly, pressed us back through Culpeper Court House, and on toward the Rapidan. At Culpeper, we had a spirited engagement in the town; as our Battery entered by one street, the enemy entered by another, and headed us off from our main course, so we turned through a back alley, and reached an open space, went "In Battery," and gave them shot at close quarters. A regiment of our cavalry charged at the same time, and checked their advance for a little while. We continued our retreat without much further molestation until we crossed the Rapidan.

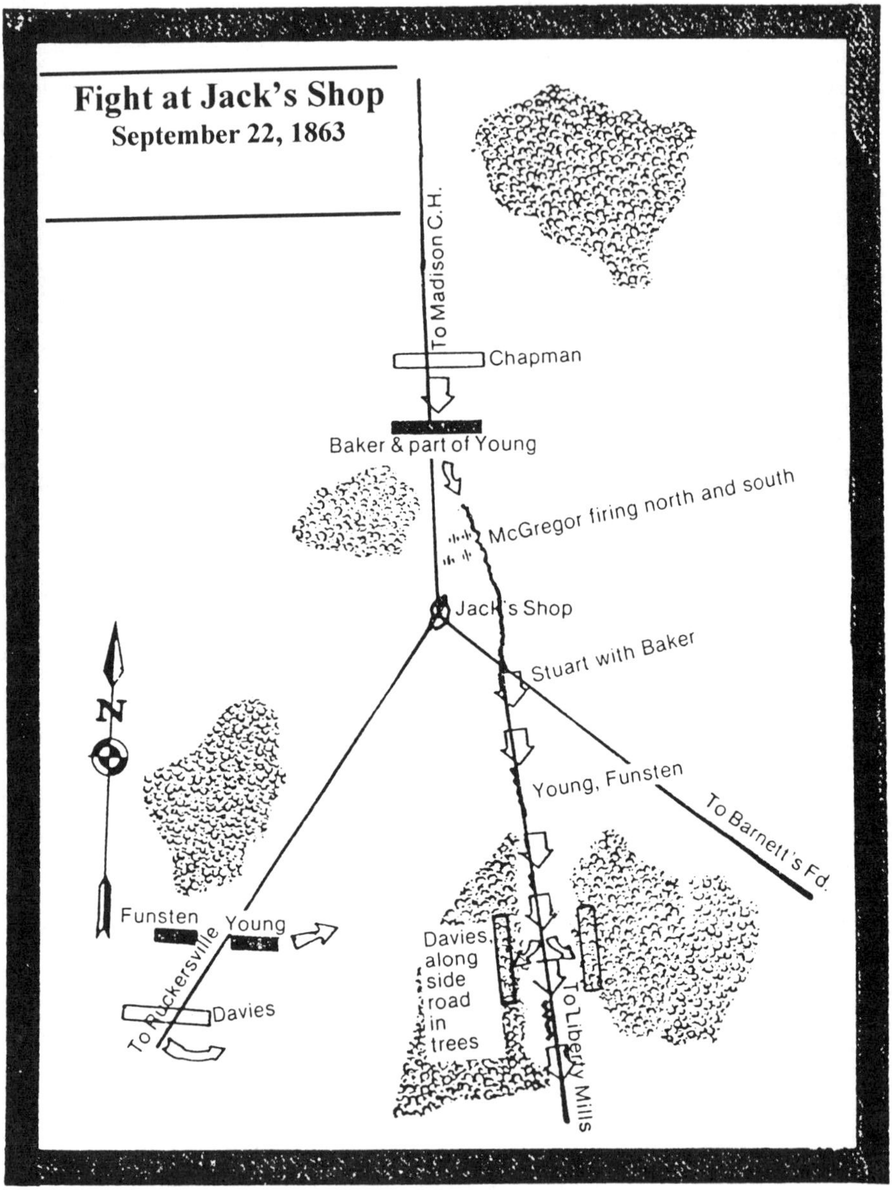
Fight at Jack's Shop
September 22, 1863
To Madison C.H.
Chapman
Baker & part of Young
McGregor firing north and south
Jack's Shop
Stuart with Baker
Young, Funsten
To Barnett's Fd.
N
Funsten
Young
To Ruckersville
Davies
Davies, along side road in trees
To Liberty Mills

Having learned through a courier that General Stuart had fallen back nearly to Warrenton, on September 19 Meade sent General Buford forward again and ran upon Stuart just five miles south of Madison Court at Jack's Shop (now called Roncelle). Buford ran into the Laurel Brigade, under Funsten, Baker's North Carolina Brigade, part of Young's Brigade, and McGregor's Battery. A Confederate charge temporarily blasted the Yankees back. Sending a dispatch to Kilpatrick, Buford saw an opportunity to strike Stuart from the rear as well as from his position, ten miles southwest of Jack's Shop. The feat was clearly possible. Stuart would know nothing of the surprise attack.

When the sound of Kilpatrick's attack could be heard, Buford again applied heavy pressure from the north. The bulk of Kilpatrick's force remained the main concern and drew the concentrated efforts of Funsten and Young. Stuart saw the criticality of the situation and exclaimed, "Boys, its fight to captivity, death or victory." Meanwhile, McGregor's guns of the 2nd Stuart Horse Artillery were sited to fire two guns to the front and two to the rear. Shreve remembered that "Gen. Stuart sat on his horse, immediately at our Battery, ordering his regiments forward and urging them to charge and clear the road in our rear, and as they passed him, they cheered and yelled and dashed forward in an irresistible charge...." After much maneuvering, Stuart worked a breakout through Kilpatrick and out along the road to Liberty Mills.

Buford pursued Stuart across the Rapidan at Liberty Mills, but was held back when he ran into Confederate reinforcements from Wilcox's Division of A. P. Hill's Corps. Meanwhile, Fitz Lee and the brigades of Chambliss, Lomax, and Owen all set out from Verdiersville to Liberty Mills to the support of their cavalry commander. Buford, being clearly outmatched, retired after eight hours of combat.

During the ongoing struggles with the enemy, for some unknown reason, on September 25, Breathed filed for a resignation. It met with strong rejections from Wickham, Fitz Lee, and Stuart. The Maryland captain remained.

Following the harrowing affair at Jack's Shop, the men of the horse artillery enjoyed the picket duty that followed for the next several weeks. Captain Breathed's Battery was again on the move on October 9, with Fitz Lee's old brigade, now under command of General Williams C. Wickham since the reorganization, to near Somerville Ford on the Rapidan. Colonel Thomas Owen, commanding the brigade in place of the injured Wickham, reported that immediately following the crossing, the enemy was encountered in line of battle near the Stringfellow house. The 1st and 3rd Virginia were ordered to charge forward, while the 4th and 2nd moved up and into the position of the charging regiments. Due to the size of the enemy force and its substantial artillery support, the men were withdrawn to the main line and dismounted along a fence line. Meanwhile, Breathed's guns were supported by those of the 1st and 3rd who remained mounted "with drawn sabers." Owen reported that the battery had been doing "good execution." With a dismounted charge of the 2nd and 4th Virginia, the enemy were driven back, and subsequently followed up by mounted charges of the 2nd, 3rd, and 4th pushing the Federals all the way to Stevensburg. The 1st Stuart Horse Artillery joined in the pursuit. Finding a good position on the hill beyond the run, the battery fired several shells into the rear of the retreating column. Losses in the battery during the affair were listed as two or three killed and ten wounded.

Moving then to Brandy Station, Owen engaged another large enemy cavalry force, but was forced to retire to Stevensburg. Joining the rest of Stuart's command, including the 2nd Stuart Horse Artillery, the combined force fought all day in driving back the enemy across the Rappahannock at Rappahannock Bridge. On October 12, Stuart continued his advance, pushing the enemy back while moving to Brandy Station, Rixeyville, across the Hazel River to Jeffersonton, and camping near midnight near Fox's Ford on the Rappahannock.

On October 13, Stuart pressed on to Warrenton, and to Catlett's Station. Near Auburn, the Confederate cavalry ran into Federal infantry. Shreve wrote that Stuart had accidentally "got hemmed in between two corps of Meade's army, who were retreating on parallel roads, and remained all night close to the enemy's bivouac, close enough to hear them talk, and see them plainly. They were unconscious of the fact that we were Confederates."

By 6:30 a.m., Stuart had been hiding his 3,000 men and seven guns for almost twelve hours. Shortly after midnight the last of French's Corps had passed through, leaving the way open for Stuart to make his way to Warrenton. However, he still waited. Having maintained contact with General Lee, the objective was to have a diversionary attack by the infantry along with Stuart's breakout. As the sun began to make its way up, a heavy fog hung low in the narrow valley beside Cedar Run. From Stuart's position, the enemy could be plainly seen preparing morning coffee and milling about their camp. With the word given, Stuart quickly ordered up all seven guns of the Stuart Horse Artillery, including both McGregor's and Breathed's batteries. Under command by Beckham, the gunners were ordered to cut the fuses exactly right. When the skirmishers opened between General Robert Rodes' Division on the Dumfries Road and the pickets of the 10th New York Infantry, Stuart gave the word for Beckham to open up.

The shell fire threw the unsuspecting infantry into an instant panic. One shell landed right in the middle of the 52nd New York and killed or

Map, Action, Auburn, Virginia, October 14, 1863

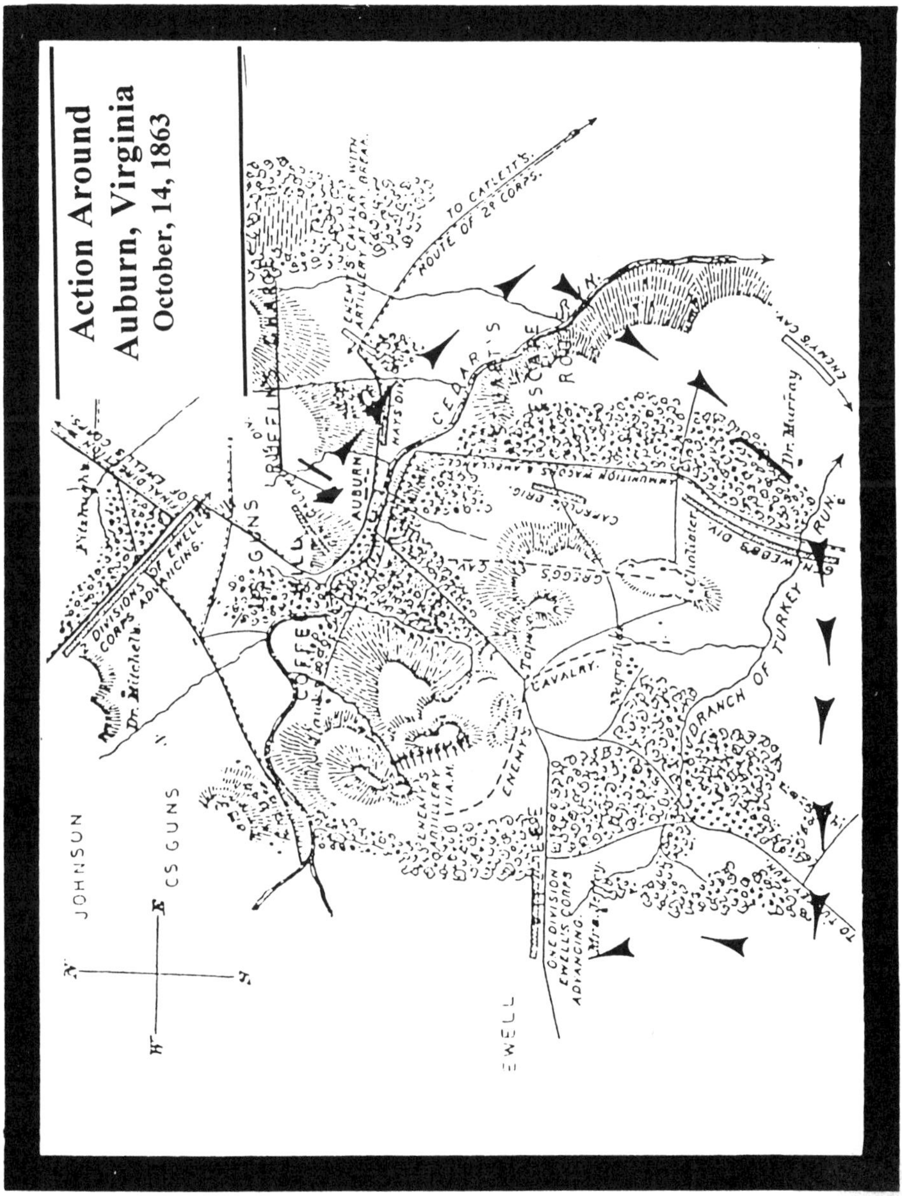

wounded seven men. With mixed commands, horses running wildly, and confusion all about, R. Bruce Rickett's battery found composure enough to return fire from atop the hill. It was soon supported by Arnold's Rhode Island Battery. The ground soon shook from the tremendous artillery fire in the low lying foggy atmosphere. Meanwhile, Joshua T. Owen's brigade of Hay's Division, not exposed to Beckham's fire, organized and prepared to move on the guns. With the 125th New York advancing skirmishers and nearly twice his number of guns against him, Beckham's guns were in severe danger. Beckham, however, did not yet know of the skirmishers approaching.

Stuart, recognizing the mismatch of guns and the precarious situation, was not going to let his favorite cannoneers down. He at once ordered the guns to limber and fall back to an unexposed position. Suddenly spying the oncoming New Yorkers approaching through the valley of Cedar Run, Stuart had Funsten's Brigade dismount as skirmishers and move against the enemy on the St. Stephen's Road. Stuart also ordered General James B. Gordon to take a portion of his brigade and charge the 125th.

After what have must have seemed like hours, the opportunity for escape finally opened up. Stuart, taking his horsemen and artillerists, rushed southward on the St. Stephens Road. Shifting then to the southwest, the horses galloped across the millrace for the flour mill at Auburn, and through lightly wooded areas before crossing part of the Old Carolina Road, just behind Federal General G. K. Warren's wagon train. Crossing the Double Poplars Road west of Auburn, Stuart finally reached Fitz Lee and Jubal Early. While the small battle continued to rage for some time, Breathed and McGregor were clear of the danger.

While McGregor's Battery gave no figures, Breathed's Battery was said to have "suffered heavily... being exposed to an enfilading fire and cross-fire from the enemy's batteries on the heights beyond Bull Run." Breathed won praise for his "splendid work." Losses from the action of October 10 through October 21 for Breathed totaled two killed and eleven wounded. Among those wounded were Martin Burke, Frank Hart, Haywood Triplett, and M. B. Young. Burke, having been wounded by a shell on October 17, had also been captured. As a result of the wound, Burke lost his ankle to amputation. Triplett lost a right leg as a result of the action at Bull Run on October 14. Young had also been wounded in the right foot and ankle but returned to serve. It was Hart who was the most fortunate. After being wounded on October 11, Hart earned a 60 day furlough to his home in Mobile, Alabama.

Barely having time to catch their breath from the Auburn incident, Stuart's cavalry again began on October 15 to screen the infantry in its railroad-destruction efforts. Moving east from Bristoe to Manassas

Junction, Stuart struck French's III Corps near Manassas. With a portion of Fitz Lee's Division in the front and dismounted, the 1st Stuart Horse Artillery and one gun from the Ashby Artillery supported the attack, forcing the enemy to withdraw. Learning of an enemy wagon train nearby, Stuart raced Gordon and Breathed's battery forward to Yates Ford. Encountering the old adversary Buford, Gordon's men, despite the support by the horse artillery, had a hard time of it striking the wagon train. Buford, determined to hold the Confederates at bay, kept the wagons rolling while defending the rear with his own horse artillery. Nightfall ended Stuart's hopes of bagging the trains.

Remaining in the area of Manassas, Stuart maneuvered his cavalry to Gainesville and skirmished frequently with the enemy along Bull Run. Lee realized however, that the area could not sustain his troops for the winter, so a withdrawal back across the Rapidan was initiated.

Prior to Fitz Lee's retirement from the area, he joined with Stuart on a last ditch effort to "punish" the aggressive enemy cavalry. Stuart marched with Hampton's division and retired down the Warrenton Turnpike. Fitz Lee moved back down the railroad to Catlett's Station, then via Auburn to a position near the Warrenton road. While under artillery fire, Lee continued to set the trap, using the sections of the 1st and 2nd Stuart Horse Artillery and Chew's single gun to counter-fire against the annoying enemy. He then dismounted his troopers and directed them into the woods on both sides of the turnpike.

Hearing the guns, Stuart turned about and struck Kilpatrick's advance guard, forcing it to reel in retreat back down the turnpike and right into Fitz's trap. Surprised and disorganized, the Federal cavalry fled through the fields to get away from the ambush, pursued still by Fitz's mounted troopers. When the Confederates captured the ford and bridge at Buckland, a large number of enemy were inevitably cut-off and captured. The pursuit of those remaining took the cavalry on a route three miles to Gainesville before the enemy reached infantry support. In the venture, horses, arms, men, and all of Kilpatrick's wagons, including Custer's papers and personal baggage, were captured. The following day, Stuart retired across the Rappahannock. The batteries made their way across Beverly Ford at 9 p.m. and went into bivouac at Dr. Green's farm.

Returning to the life of pickets, the 1st and 2nd Stuart Horse Artillery enjoyed the intermission from war temporarily. On November 1, the 2nd Stuart Horse Artillery, according to Shreve: "had an engagement at the railroad bridge crossing the Rappahannock, where we did some good shots at the enemy on the north side, but no result of interest. The 1st Stuart Horse temporarily lost the services of Lieutenant Shanks when he was wounded on November 2."

Map, Buckland Races, October, 1863

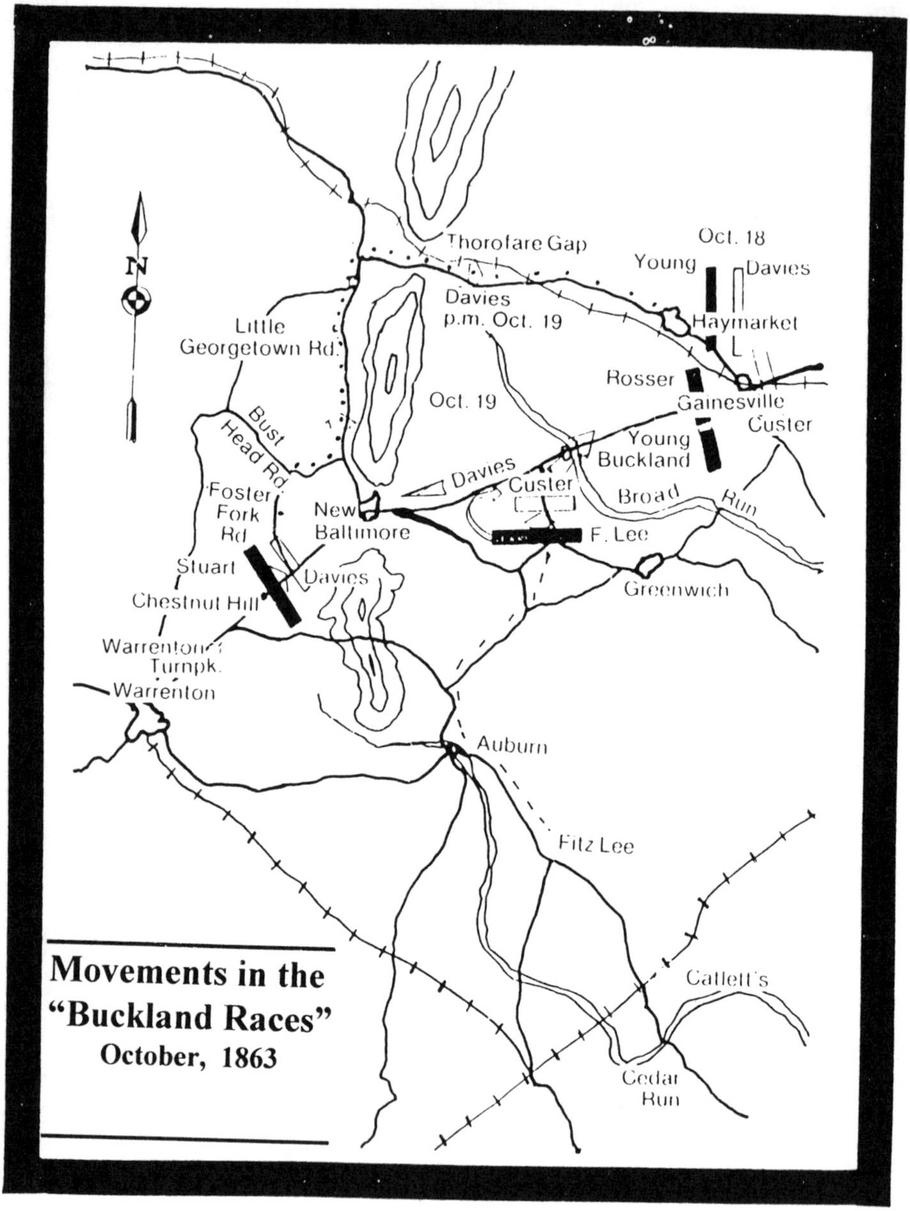

On November 5, Stuart held a grand review of cavalry at Botts Farm with General R. E. Lee as the guest of honor and Governor John Letcher. The ruckus that was made from the review, not unlike the review of June, caused a stir from the enemy, but ended in no major engagements. Following the advance of the enemy across Kelly's Ford, on November 8 Stuart sent out his cavalry to Brandy Station. Arriving at Fleetwood Hill, several guns of the horse artillery battalion went on line only to move again to the Kennedy Farm between Brandy Station and Culpeper Court House. Shortly after unlimbering the guns, the cavalry moved back on the same line prior to the advance of the Federal cavalry. Upon the appearance of the foe, the Confederate batteries opened and checked the advance long enough for Federal artillery to arrive and return the shells. Moorman's Battery held the post until Breathed's and McGregor's batteries formed a short distance in the rear, providing the Lynchburg gunners with a secondary line of defense.

Moorman's Battery was soon after ordered to the left on the Rixeyville road, followed by two Napoleons of McGregor's Battery. Sleeping at the guns that night, the batteries were involved with the enemy the following day and aided in driving the Federals from the field. Only Patrick White of the 1st Stuart Horse was recorded as wounded during the fighting around Brandy Station on November 8.

On November 7, both the 1st and 2nd Stuart Horse were again engaged. In an effort to hold off a strong enemy force of artillery, cavalry, and infantry, the cavalry and horse artillery was rushed toward Brandy Station. First taking position on Fleetwood Hill, the battalion was later moved to "a position on the road between Brandy Station and Culpeper Court House, on the farm of Mr. Kennedy, to prevent them from flanking or getting in the rear of our forces at Fleetwood Hill." Initially, Shoemaker's guns took the first shots on the enemy until fired on by the opposing artillery. Holding under a heavy fire, the Lynchburg men were relieved when cavalry and the 1st and 2nd Stuart Horse Artillery formed just a half a mile to the rear, allowing Shoemaker's guns to retire. The only thing that Shreve recalled of the day was that at Kelly's Ford "quite a body of our Infantry suffered a miserable defeat and surrendered."

Throughout the movements of the year, Battalion Ordnance Sergeant Walter S. Dabney had made sure that the battalion was well supplied. Surviving requisitions are sparse, but show a large usage of ordnance from June until December 1863. On June 12 Dabney had received sixty rounds of Blakely, 36 canister and 160 friction primers. On August 31 he received 16 rounds of 12-pounder shot, 16 of shell, 24 canister, 24 case, and 120 friction primers. August 31 lists the highest draw of ammunition with 1,140 3-inch rifle shell, 492 12-pounder Napoleon shell, eight Napoleon canister, 32 Whitworth shells, one caisson and limber, three 3-inch rifle sponges, 150 Colt Army and 36 Navy pistol cartridges, and 100 pistol caps.

On October 15 the demand had declined, with only 160 rounds of 3-inch shell requested, and 240 friction primers.

Soon thereafter, the batteries of the Stuart Horse Artillery battalion were called upon to join Fitz Lee's Division and others in the Shenandoah Valley to intercept raiders. However, after moving as far as the base of the Blue Ridge by November 18, the alarm was called off with the raiders withdrawal from the Valley. Returning to the vicinity of Orange Court House, the Ashby, Lynchburg, and 1st and 2nd Stuart Horse Artillery all went into camp near Verdiersville on the road leading from Fredericksburg. While the weather grew significantly colder, the companies never had an opportunity to build winter quarters, being removed again to the hills near Orange Court House, and then again moved to the road northwest of that town. The 2nd Stuart Horse Artillery recorded that it was near Milford Ford on November 23. Picket duty was resumed, and on November 26, a member of Moorman's Battery recalled that it was the "coldest we have had this winter."

The night before, Meade had shattered any hopes of the Stuart Horse Artillery battalion for winter quarters as he put his army in motion across the Rapidan at Germanna and Ely's Fords. Later on November 26, the men of the horse artillery again found themselves on the move. The batteries arrived in the vicinity of Mine Run on November 28.

Taking various positions along Lee's Mine Run line, some batteries became engaged briefly. However, on December 3, the batteries were pulled off the line and marched back to the vicinity of Orange Court House. There, the men reluctantly began to build winter quarters, knowing that they would yet again be called away. In fact, the batteries were again ordered to move and found themselves on the march to Charlottesville on December 21. Private "Davy" Cardwell of the 2nd Stuart Horse Artillery remembered:

> As the command reached the Rivanna River, the lead team took the water and proceeded to the other side, but when they attempted to pull the gun up the bank of the river the horses' feet slipped on the ice which had formed from the drippings from the lead horses, and they could not budge the gun, which was yet in the river and which the river was from three to four feet deep. The horses were whipped and pushed and the men shouted, but all to no purpose; the horses could not get a footing. When this was demonstrated, Capt. Wilmer Brown turned in his saddle and gave the command: "Cannoneers, dismount; by hand to the front!" Now, there you are; get down into the icy water and put your shoulders to the wheel. Cold? Cold was no name for it. But down we got, up to our waists in the ice water...

We finally had to attach prolonged ropes to the tongues of the limbers and pull them out separately by main strength of the men. Howard Neal of the 1st Stuart Horse Artillery was the only man listed as captured during Mine Run.

For the next several days, men worked hard in harsh cold weather to complete warm dwellings for themselves and their horses. The place was described by H. H. Matthews of Breathed's Battery as "a beautiful forest near the road leading from Charlottesville to Madison Courthouse, crossing the Rivanna River near the Rio Grist Mill, and distant from Charlottesville northeast about three miles." John F. Key, 1st Stuart Horse, was listed as captured at Madison Court House on December 28. Matthews also recalled:

> We immediately became very busy building very comfortable houses (or shacks, in army parlance), topped off with tent flags and stick chimneys chinked with Albemarle mud. This was the first time in the experience of the battalion that we had been so comfortably fixed. Our shacks were so arranged that they could hold six men. Bunks were built one above the other, two men sleeping together in the two lower bunks and one man each in the upper bunk, thus accommodating six men without too much crowding. We chinked the logs with mud and chips, excluding all air, making ourselves as comfortable as a "bug in a rug."

Matthews also spoke fondly of the trusted horses of the batteries and of the need to harbor them as well as the men:

> We also cut down immense oak trees, making them into shingles and lapboards to cover the frame work of the stables that we had erected for our faithful old battery horses that had passed through that fearful campaign of the spring of 1863... How dear those old horses were to us! It seemed as if they were verily a part of us; that they knew every man in the battery who unconsciously had contributed to their misery and burdens. Many night have we started out on an expedition that had for its object the securing of forage for those famished horses. Many an unsuspecting farmer was visited during the night for that purpose. It was not our purpose to willingly neglect any of our farmer friends in our nocturnal visits, but treat all alike and in the morning when the farmer would begin to take stock of his feed, corn and fodder, he could not help exclaiming: "Who's been here since I went to bed?" And in the same breath would answer: "That Horse Artillery Battalion, God

darn them!" Our camp would, of course, be visited by the indignant granger, bent on finding the Midnight marauders, but never finding anything but a lot of reckless Johnny Rebs, who would laugh at the indignant manner in which Captain Breathed would dismiss the aforesaid granger with the most emphatic assurance that the men of his battery represented the best blood of the good state of Maryland; that they were gentlemen, and did not do such naughty things. At the same time Old Bill, Captain Breathed's faithful old horse, was quietly nibbling on some of that identical fodder. I myself plead guilty to this breach of hospitality, and throw myself upon the tender mercies of the farmers of dear old Albemarle.

This time, the winter quarters site was genuine.

Chapter 5

1864

"....whenever or where ever there is likely to be a fight, they are bound to be in the front."

The first day of the year blew in quietly for the battalion in winter quarters. The men all took pleasure in finishing the stables despite heavy snow. Until February, little of importance took place with the exception of the daily picket at Rio Bridge on the Rivanna River. The men of the horse artillery took the rare opportunity to relax in peace for what they hoped would last at least until the spring.

Quartermaster Matthews, in later years, remembered of the time:

> We had a most delightful camp, and made a number of acquaintances among the people of the neighborhood and of Charlottesville. I only regret that I cannot recall the names of some of the good people that Helped in many ways to make our time pass pleasantly.... I was Particularly fortunate in the possession of some dear friends whom I had known in Georgetown, D.C. at the commencement of the war, who went to various places in the South, finally landing at the Parish House and in the town of Charlottesville. I spent many a pleasant hour with them during our encampment near Charlottesville, sitting at their table every Sunday, enjoying rye coffee and other well-cooked food that to a soldier seemed like angel food indeed.

Captain McGregor's file in the service records shows a great deal of supply activity during this time. It was time to resupply the men with clothing items.

The 1st Stuart Horse lost a few officers as a result of commission and transfer in February.

Major Beckham was transferred to the western theater on February 16. Promoted colonel and given command of the Army of Tennessee artillery, the post must have seemed a tremendous boost in his rise toward a generalship. The transfer would, however, leave a void in the high command of the Stuart Horse Artillery for only the second time since its organization and would permit another to rise to the head of the famous organization. However, within nine months of his transfer, Beckham would lie among the mortally wounded at Franklin, Tennessee, a victim of the terrible battle on November 29. Surviving for less than a week, Beckham was laid to rest in St. John's Churchyard in Ashwood, Tennessee.

William T. Mason had been assigned to a position as aid to his brother-in-law, General E. P. Alexander. When speaking of "Willie," Alexander wrote: "He was my wife's youngest brother - was only 17 years of age, & was a private in Breathed's battery of horse artillery, in which he had distinguished himself for his coolness & gallantry, when I applied for him as my aid. He was very young-looking, even for his age, & he became a sort of favorite with the infantry of the corps, who got to know him & would frequently call after him on the road, 'There goes Lieutenant Willie.'" Mason later distinguished himself again in manning the guns at the Crater in Petersburg and by burning the Farmville bridge in April 1865.

As was typical with the Stuart Horse Artillery Battalion, however, war was always a short distance away. The Confederate winter quarters routine was interrupted when General Kilpatrick moved to attack Richmond. As a decoy, Colonel Dahlgren was sent to the west of the city to demonstrate and draw away the Confederates in the area and thus weaken the line in the east. As soon as Kilpatrick was found to have broken through the lines at Spotsylvania Court House, General Wade Hampton, being encamped at Bowling Green in Caroline County, gathered about 250 men of the 1st North Carolina Cavalry, and 40 men from the 2nd North Carolina Cavalry for the pursuit. Sometime in the rapid organization, a section from McGregor's Battery, under the command of Lieutenant Edward Sully, was dispatched from Albemarle to assist.

Leaving camp at about midnight on February 29, the contingent had marched through a terrible winter storm of rain, hail, sleet, and snow for nearly 24 hours when they arrived near Atlee's Station. Daring to make a night attack, Hampton deployed several of the Tarheels on foot and stationed one gun in support of the advance. With the opening gun, the 120 men of the 1st North Carolina attacked the sleeping Federal camp, taking it completely by surprise and capturing more than 150 prisoners, 80 horses, carbines, sabers, saddles, bridles, blankets and numerous other items. The brigade that Hampton attacked turned out to be the rear of Kilpatrick's column. While a few scavenging horse soldier's gathered the various articles of war, the remainder of the attacking force pursued the rest of the Federal brigade to the camp of the main body. Sully's lone gun continued to move and shell the Federals along the road. By morning the demoralized Federals were en route to the Pamunkey.

While Sully was moving with Hampton from his camps in his attempt to cut off Dahlgren, just after midnight on February 29, General George Custer and 1,500 Federal cavalrymen mounted at Madison Court House and prepared march southwest along the Rapidan River. Accompanying the contingent was the horse battery of Company E, 1st U.S. Artillery, under Porter. Upon approaching Wolftown, the Federals met a picket of Fitz Lee's Brigade posted north of Bank's Ford on the Rapidan. The Southern pickets, being poorly supported, fled at the sight of the enemy.

By daylight, Custer's force crossed and pressed on to Stanardsville, where it arrived by 8:15 a.m. Again, Confederate pickets there quickly fled in the direction of Jack's Shop.

News of approaching enemy cavalry arrived in the winter camp of the horse artillery between 10 and 11 a.m. However, initial reports that the enemy was within six miles were discounted as exaggerations or amateurish. Winter camp life continued with little concern. Having just arrived from the mail run to Charlottesville, Mark Fuqua and a few others of Shoemaker's Battery were ordered to remount and return to the river to see what truth there was to the reports. Major Moorman initially headed out with them, but upon passing Dr. Cooke's home, stated that he had to return to unfinished business in camp. Within ten minutes the contingent was captured. Moorman, upon returning to camp at about 12:30 was met by Lieutenant James Cunningham of Company B, 1st Virginia Cavalry. Cunningham's report was no different than those received earlier in the morning; except for the fact that the enemy force was even closer. The winter camp was at once ordered into a frenzy of activity.

More pickets were hurriedly sent up to Rio Bridge, one and a half miles to the north, to ascertain the number of enemy advancing; however, upon their arrival, the Federals moving along the Earlysville Road, were already crossing in force at 3 p.m. At the same time another column was discovered to be crossing at Cook's Ford, almost two miles below.

The Confederates whom Custer's "legion" had captured gave false intelligence about the number of Confederates to the front. Fuqua recalled that they were questioned on the south side of the river on the hill from the bridge at Berkeley's Mills. In his post-war recollections he stated that Custer himself had interrogated him and Littleton Moorman. In no time the actual encounter of forces had begun, Custer's large force faced nothing more than the battalion of Stuart Horse Artillery. However, Custer believed that he was facing much more, including a superior force of cavalry supported by infantry.

Having gone to the river himself to see the enemy advancing, Captain Breathed returned hurriedly to camp and ordered Lieutenant P. P. Johnston out on the road with his section of guns. Matthews remembered that "this was done quickly, but not too soon; for as soon as the guns were placed in position, the enemy came around a slight bend in the road from the direction of Rio Mills. Breathed opened on them with both guns, double shotted with canister. This had the effect of causing Mr. Custer to stop and think and wonder why he had been so unceremoniously stopped in pursuit of his pet project, but he did stop."

Shreve of the 2nd Stuart Horse Artillery recalled, "We wheeled some of the guns into position and opened fire on them, which checked their forward column, and during the short respite, we got the horses hitched and formed the other guns in line...."

Matthews soon heard "Fours, left about." The guns had temporarily held back Custer, but the men knew they had not seen the last of him yet. Reforming and throwing out skirmishers, Custer's men renewed the advance. Again Matthews:

> Lieutenant Van Shanks, who commanded the other section of the battery, as soon as he could get his horses in the guns moved rapidly to Breathed's aid, as did Captain Preston Chew... As soon as Shanks relieved Breathed, Breathed moved his pieces by hand to the rear and hitched his horses to them, taking position to the rear of Shanks, who retired to the hill that Moorman had moved his other battery to... A shell from Breathed's No. 1 gun exploded at the head of Custer's column, killing and wounding several men and horses. This caused Custer to halt and throw out flankers in order that he might discover our strength.

The few cannon of the Stuart Horse that were prepared gave fire. Lieutenant Phelps of Moorman's battery recollected that "men and horses were running everyway.... I had one of my pieces put in position and commenced firing left and right." Meanwhile, the Federal cavalry which had crossed at Cook's Ford (Captain Ash of the 5th U.S. with two squadrons of the regiment, about 65 men in all) charged "wildly" into the Confederate quarters, first into Chew's portion, cutting down caissons and scattering fires in the abandoned huts and stables.

Leaving most of the caissons and baggage in the "care" of the enemy, the batteries did succeed in getting the guns off to an elevated position and opened into their own winter home. Breathed reportedly shouted out to his men at their guns in the road to "stand to your guns boys! You promised me you would never let these guns be taken." While the guns were served with the fewest men possible, Breathed and Chew formed the remainder of the men to the rear of the guns in an effort to build a "cavalry" force of some nature. With not one rifle or carbine and a limited number of pistols, sticks, clubs and even fence rails the men formed in a line as a show of force. As the enemy wavered and began to withdraw from the force of the cannon fire, Breathed's and Chew's "cavalry" dashed out yelling out false commands to the rear to indicate that they were only a portion of the force present. Chew reportedly called out "Tell Colonel Dulany to bring up the Seventh Regiment!" The charge struck Ash's squadrons and threw them back across the river, and continued the pursuit for some distance while Shoemaker continued to throw shells in on the retreating column. Additionally, train whistles nearby could not have been better timed. While they were later convinced that no Confederate infantry was at Charlottesville prior to their raid, Lieutenant George Yates, aide-de-camp to Pleasonton, reported that the infantry "had evidently just arrived by trains." Combined with the horse artillery sham, Custer was convinced he was facing a superior force and withdrew that evening to a position southwest of Stanardsville.

Custer's expedition into Albemarle had cost the battalion of horse artillery a great deal, but the guns of the battalion had been saved. Matthews, along with several other men had lost several items in the fight: "I was a loser in a pan of light bread that I had made that morning, and fully expected to eat for supper; but man proposes and God disposes, was beautifully exemplified in my case." Moorman, in the wake of the event at Rio, and in command of the battalion when it was attacked, reported the losses in the aftermath:

> Moorman's Battery, two men and two horses captured; Chew's Battery, ten sets of harness, one limber with canteens, one forge, six tents, five tent flies, four tarpaulins, sixty pounds of axle grease, fifteen curry combs and brushes, three public horses, forty government bags; Breathed's Battery, nine tents, two horses, three and one-half sets of harness; McGregor's Battery, six sets of harness, three tent flies, twelve bridles, six saddles and blankets, four halters, two mules, four skillets, four water buckets. In addition Breathed's and Chew's men lost heavily in private effects.

Upon recrossing the river Custer saw to it that the bridge and nearby flour mill were also burned. Captain Ash's group came out with only one man slightly wounded. Matthews recalled that Stuart reached them "about the time Custer was going into Madison county." He continued of the rest of the miserable evening:

> It began to rain about dark, one of those cold, freezing March rains, that would make a fellow as if he wished the Yankees were in the infernal regions, and I in my father's library sitting by a red hot stove. As the rain fell it turned to sleet. I do not remember ever having experienced a more disagreeable day. Stuart passed us about dark and met the

enemy's rear guard at Stanardsville, and had a slight engagement.

On March 3, an alarm again rose through the camp that the Federals were closing on Charlottesville. In an effort to move more quickly than the enemy this time, the men at once hitched up the teams and individual mounts. The alarm however was no more than a falsehood, leaving the cannoneers to return wearily to their desolate quarters. The following day the same alarm was sounded, sending the battalion out upon the Charlottesville road only to be recalled as the day before.

On March 20 the winter encampment of the battalion was abandoned as orders arrived to move to Gordonsville to camp and wait for the spring campaign. By 10 a.m., the men, horses and guns were en route, moving through Charlottesville and crossing the Rivanna until coming to a halt in the evening within ten miles of the destination. McVicar, of the Ashby Artillery, commented in his diary that there were "plenty of ground hogs here." The following day the battalion reached Gordonsville and resumed winter life for the next few weeks among weather filled with snow and sleet. After having left their "beautiful camp" in Albemarle, Matthews recalled the heartbreak of the move and the disappointment on arriving at Gordonsville: "bidding good-by to many friends, nevermore to look on that spot again....to the mudholes of the vicinity of Gordonsville where they received new horses and equipped generally, as a preparation for that never-to-be-forgotten campaign of 1864...."

On April 22, in appreciation of the efforts at Rio Hill, the ladies of Charlottesville made a presentation to the Stuart Horse Artillery. The gift was a large silk flag on which were inscribed the words, "To Our Brave Defenders, the Stuart Horse Artillery." The kind token would be a matter of honor and pride in years to come. It would travel in the postwar years with its final bearer, Nimrod Ware, to the Far East until finally coming to rest in the hands of Colonel Chew in Charles Town. Some time later the flag was put in the Handley Library in Winchester before being transferred to the Jefferson County Museum, where it rests today in a fine glass frame.

A resolution that the Charlottesville town council passed read:

> Whereas, The recently attempted raid of the Yankees on this place was undoubtedly checked and finally repulsed by unequaled coolness and courage of the gallant officers and men of the artillery battalion encamped a few miles north of Charlottesville, wholly unsupported as they were by either infantry or cavalry; and

Whereas, Our town was thus unquestionably saved from pillage and the public stores and the railroad bridges from destruction; therefore, be it

Resolved, That on behalf of the citizens of Charlottesville, we, the council of the town, do hereby return our thanks to the officers and men of the said artillery battalion for their gallant and heroic conduct on the occasion above mentioned, with the assurance of our lasting and grateful appreciation of the service thus rendered us.

Resolved, That the above preamble and resolutions be handed to the officer in command of the battalion in order that he may communicate the same to the officers and men of his command in the manner he may deem most appropriate.

By order of the council, March 7, 1864.

A. Robert McKee, Clerk.

Returning to the reality of life in the field, a rare artillery practice shoot was held on April 26 by the battalion. The event was so unusual, that George Neese, of the Ashby Battery, recollected it as the first in his memory since the beginning of the war.

Prior to and after the arrival at Gordonsville, there was a great deal of reorganization within the Stuart Horse Artillery Battalion. Perhaps the most significant, in the wake of the recent transfer of Beckham to the west, was the promotion of James Griffin Dearing, Jr., to lieutenant colonel and command of the battalion. Dearing, a native of Campbell County, had, like the two previous commanders of the horse artillery, attended West Point. Dearing had resigned in April 1861 at the outset of the war. First holding a commission as 2nd lieutenant with the Washington Artillery of New Orleans, Dearing later rose, on April 3, 1862, to command what had been Henry G. Latham's Lynchburg Artillery. Promoted to major and overall command of the 38th Battalion Virginia Light Artillery by May 1863, Dearing continued to gain recognition as an excellent artillery officer. Due to this recognition, Dearing's star continued to rise, being assigned to command of the cavalry in the district of North Carolina during the winter of 1863. Finally appointed lieutenant colonel on March 14, 1864, Dearing attained the command of the horse artillery for the Army of Northern Virginia. This command was short lived, however, as he continued activities in North Carolina. Dearing's fame led him to a brigadier generalship dated April 29, 1864, followed by his untimely death at High Bridge on April 6, 1865.

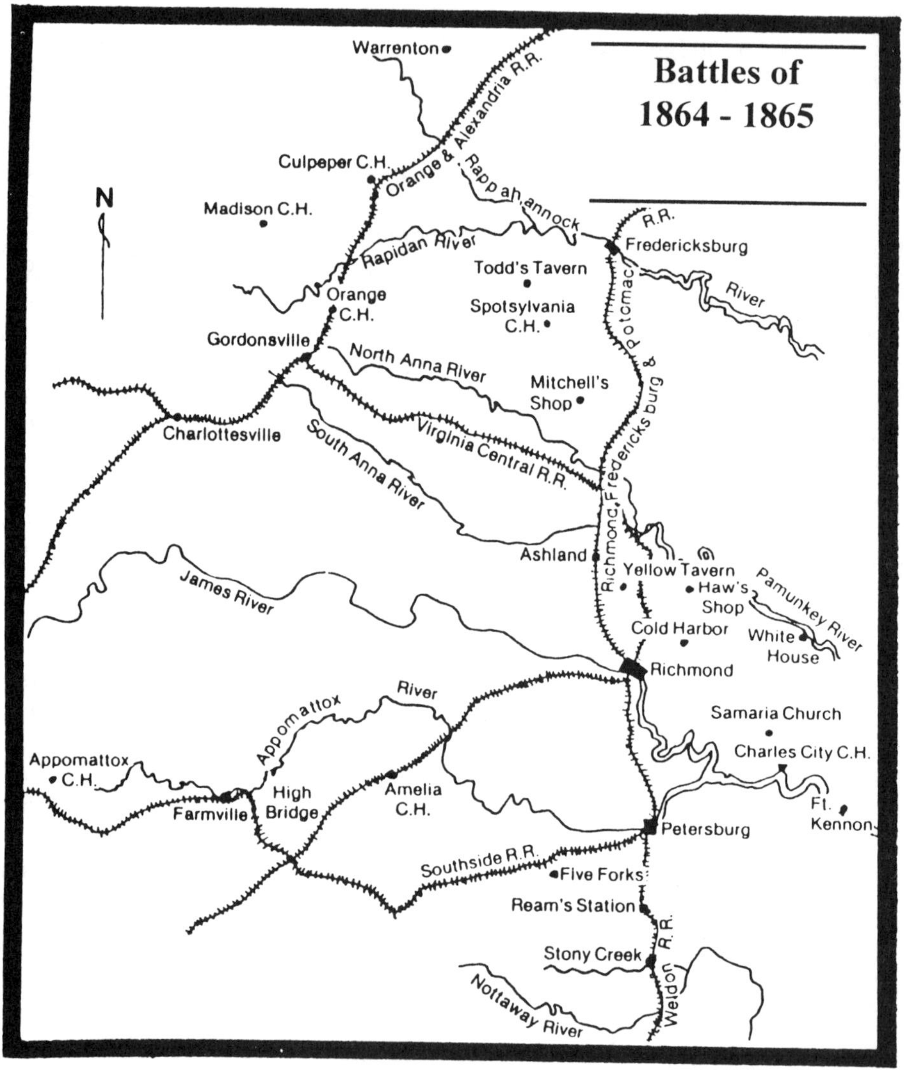
Battles of
1864 - 1865
N
Warrenton
Culpeper C.H.
Orange & Alexandria R.R.
Rappahannock
Madison C.H.
R.R.
Fredericksburg
Rapidan River
Todd's Tavern
River
Orange
C.H.
Spotsylvania
C.H.
Potomac
Gordonsville
North Anna River
Mitchell's
Shop
Charlottesville
South Anna River
Virginia Central R.R.
Richmond, Fredericksburg &
Ashland
Yellow Tavern
Haw's
Shop
Pamunkey River
James River
Cold Harbor
White
House
Richmond
River
Appomattox
Samaria Church
Appomattox
C.H.
Charles City C.H.
High
Bridge
Amelia
C.H.
Farmville
Petersburg
Ft.
Kennon
Southside R.R.
Five Forks
Ream's Station
Stony Creek
Weldon R.R.
Nottaway River

Dearing really never held physical command of the battalion. It was Roger Preston Chew who occupied the post officially and held it to the close of the war. Chew, promoted to major and second in command of the horse artillery on February 27, 1864, had quite a career as a horse artillerist. The organizer of the first horse artillery company in Virginia and the Confederate service in November 1861, Chew was the forerunner of the reintroduction of horse artillery during the Civil War. Chew had attended the Virginia Military Institute and had been one of the many cadets to act as drill masters in Richmond in April 1861. Temporarily assigned to the Lee Lynchburg Battery at Monterey, Virginia, in the summer of 1861, Chew was officially appointed a lieutenant of artillery and assigned to serve with Brigadier General Edward Johnson. Chew's class was graduated early, in December 1861, though in fact he had been in the field since April. On November 13, 1861, Chew formed the Ashby Battery of horse artillery, which served with distinction in the Shenandoah Valley Campaign of 1862, though he was not officially appointed to captain until April 1862. Though Chew did not see an official promotion to lieutenant colonel until February 1865, he would command the Stuart Horse Artillery battalion organization.

Structurally, the horse artillery batteries suffered through much confusion in reorganization. With the loss of Dearing as potential commander, the battalion was to be officially reorganized again in late April. With the promotions of both Chew and Breathed, the organization was supposed to have two "demi-battalions." However, after the loss of Dearing as battalion commander, the organization was restructured with Chew as senior major commanding the battalion, and Breathed as major and executive officer. Though the 1st Stuart Horse Artillery would lose its captain to promotion, he would never be too far away as the battalion's second in command. In the lower echelons of "officer country," "Dick" Shreve received his promotion to 1st lieutenant in the 2nd Stuart Horse as a result of a recommendation by Stuart himself.

On May 3, General Stuart performed an inspection of the horse artillery battalion. It left Gordonsville to join the main army the following day. In the early morning of the 5th, Johnston's, Shoemaker's, and Thomson's batteries reached Shady Grove, where the battalion went into action. One of Shoemaker's guns and two rifled pieces of Thomson's Battery were recorded as participating in the action, but there is no record of the 1st Stuart Horse Artillery's participation.

General Ulysses S. Grant had sent the Army of the Potomac across the Rapidan at Ely's and Germanna Fords, forcing the Army of Northern Virginia into action. In a dispatch early on May 5 to Major H. B. McClellan, Chew stated that somewhere behind the batteries would follow the trains, both ordnance and quartermaster. Chew wrote that "MacGregor will equip his battery with horses from the lot expected last night, and has orders to report to Maj. Gen. W. H. F. Lee, by daylight, at Orange Court-House." Though few details survive, Johnston's Battery was engaged near Shady Grove on May 6.

It was also during this time that transfer of the 1st Stuart Horse Artillery was considered. The unit was composed primarily of Marylanders, so the question was raised whether the battery should not actually be transferred to the Maryland Line. Even during the rapid movements and active participation of Johnston's Battery in the days around the Wilderness, General Stuart took the time to inform General Samuel Cooper, Adjutant and Inspector General, that he did not favor the transfer of the battery to the Maryland Line. In his letter he went to great lengths to protest the proposed action:

> This [Breathed's] battery as well as McGregor's was, during the first year of the war, organized under the auspices of the Governor of Virginia as part of the State troops, and enlisted for the war, the officers receiving their commissions from the Governor of the State. Under the immortal Pelham it received accessions by recruits from Maryland, and while Captain Breathed remained with it, 3 of its 4 officers were native of the State.... But the most weighty reason why this transfer should not be consummated is that at the very commencement of a momentous campaign a battery which has won for itself a name second to none in this army, whose services cannot be dispensed without great injury to the cause, will be disorganized and rendered almost entirely useless, and this great injury to the battalion of horse artillery will be without any corresponding gain to our service in any other direction, for these men will probably remain in a camp of instruction for a considerable portion of the coming campaign, and in lieu of the distinguished service they might render in the field, will consume their time in the monotonous duties of the camp. I visited this battery yesterday, and find that many of the men are changing their opinions, and, desiring to remain in that company, are requesting that the order for their transfer to be revoked.

While a few individuals would hold firm to their decisions to transfer, the company itself continued to maintain its Maryland compliment as a Virginia organization. On June 9, the order was recommended for suspension "if not unconditionally at least for the present.... In this case it would really amount to a disbandment of the company. This is an inopportune time for such changes."

For all intents and purposes, by the second week in May, 1864, the 1st and 2nd Stuart Horse Artillery went their separate ways. Attached to different commands and beginning to operate in different theaters of action in Virginia, the association between the two would be a matter of history.

Chapter 6
May - December
1864
A Parting of Company:
The Split of the 1st and 2nd Stuart Horse Artillery

The 1st Stuart Horse Artillery

Attached to Lunsford Lomax's Brigade, on May 8, the 1st Stuart Horse Artillery again came under fire near Spotsylvania Court House. As Federals from General G. K. Warren's V Corps pressed forward, Lomax retired from one wooded rise to another about a half-mile farther on, just to the northwest of the intersection of the road from Piney Branch Church and the Brock Road. It was in this improved position that Lomax decided to make a brief stand. Overseeing Johnston's four-gun battery was Breathed himself.

Having retired to a clearing at the Alsop farm, the cavalry and Johnston's gunners prepared for the oncoming foe. Atop a slight eminence along the road, Fitz Lee detached one section of Johnston's Battery and a few dismounted skirmishers to delay the advance so the remainder of Lomax's Brigade could continue the withdrawal to a better position. Lee and Breathed remained at the guns. As the enemy closed in, the two guns unleashed a fire that threw the Federal infantry into disorganization in its column formations and forced them into battle line. With the 13th Massachusetts in the lead, the regimental colors were brought up and the regiment advanced with a rush toward the guns. A shot from one of the guns succeeded in breaking the flagstaff and wounding the color bearer, but did not hinder the Bay State men as they continued toward the prizes. Seeing disaster imminent, the Confederate skirmishers fled, leaving Lee to order Breathed to bring off the guns.

In the ensuing action, the battery commander was hit and the onrushing Massachusetts men were so close as to begin yelling for the battery to surrender. Ordering the last piece to limber and retire, Breathed urged it off, though only one driver remained mounted. Other crew members had already fled or been wounded. In a parting volley from the Federals, the last driver's arm was shattered. Seeing the situation in peril, Breathed leaped from his own mount onto one of the wheel horses to drive the gun off. No sooner had Breathed mounted than a lead horse tumbled to the ground wounded. Dismounting, the major rushed to cut the horse from the harness. As soon as he cut the one loose, another fell, forcing more delay while he ran to cut loose the other one. Finally, Breathed mounted the wheel horse and was successful in getting the piece out of harm's way.

On the following day, General Philip Sheridan with his Federal cavalry moved along the Telegraph Road toward Richmond. While only held back initially by Wickham's Brigade, Sheridan was temporarily kept in check when Fitz Lee came on the scene with reinforcements, including Johnston's and Griffin's batteries and a section from Hart's. The attempt to stop Sheridan failed at Beaver Dam Station. Stuart personally rushed forward to Yellow Tavern with his cavalry to stop the enemy raid. With Wickham posted on the right and Lomax on the left, Major Breathed saw to the positioning of his artillery. However, owing to the nature of the terrain, placement was limited. One single piece from Hart's Battery occupied the road while Johnston's Battery was posted on an elevation just to the rear of the cavalry line.

Early on May 11 Sheridan struck Stuart's lines with devastating effect, carrying off several prisoners and horses from Griffin's 2nd Baltimore Battery. The guns of the horse battery were untouched, however. Later in the day, Federal cavalry charged the position of the 1st Stuart Horse Artillery; but between the artillery fire and the efforts of the 1st Virginia they were driven back in disorder. John Esten Cooke remembered of the attack:

> Stuart was beside Breathed's guns. It was difficult to use his artillery in such a melee of friend & foe & his left wing was soon in utter disorder. The federal attack had at last succeeded in breaking to pieces, the men were scattering in every direction, & Stuart shouted to Breathed, "Breathed take command of all the mounted men in the road, & hold it against whatever comes. If this road is lost we are gone." At Stuart's orders Breathed saluted & shouting to the men to follow charged the Federal column. He was immediately surrounded. He fell in the enemies [sic] line. Instantly he realized his situation, a sabre blow cut him out of the saddle in the Federal line. The first opportunity he had was seeing an officer riding by he gave a spring & _____ his horse, mounted & made his way out, he distinctly hearing the shout Kill, Kill him! So near him was the Confederate line, men shouting - "God sake boys let us go back they've got Breathed."

While few other details are available to indicate the role of Johnston's Battery in the episode, Hart's guns were recorded as being hotly engaged. By the close of the day, the Confederate cavalry had been forced to retire from the field, Breathed retiring with them. Major Breathed was lightly wounded, but the heaviest cost of the engagement had been the mortal wounding of the beloved cavalry commander. Stuart, the "Beau Sabreur" and founder of the original Stuart Horse Artillery battery, died in Richmond on May 12.

The 1st Stuart Horse Artillery continued with Fitz Lee's Cavalry Division. While Lee marched his division toward Mechanicsville in an effort to resupply and continue the monitoring of Sheridan, he penned an official announcement:

Hd. Quarters Lee's Cav. Div.
May 14, 1864.

General Orders
No. 6

A terrible duty has to be performed in announcing to this Division the death of Major General J. E. B. Stuart, who died in Richmond on the evening of the 12th from the effects of a wound received in the engagement near Yellow Tavern on the 11th inst.

Associated with the Cavalry of the Army of Northern Va. from the commencement of our struggle - his name and fame, bright as the keen blade of his trusty sabre, is at this moment encompassed by no boundaries. Great, glorious and good - his loss to his Country - to our Army - Especially to his troops, is inconsolable.

Whilst his bright glancing eye can no longer see - his clear, ringing voice no longer heard by his mourning followers - may the principles he has taught us - the example he has shown us, not be lost. Stuart had no superior as a soldier. He fell on the battle field, fighting with dash, daring and enthusiasm, which strongly characterized him, and it became the fortune of the gallant Old 1st Virginia, his Old Regiment, to rescue him, bleeding but undaunted from the foul touch of the dastard foe.

Let the remembrance of his mighty deeds, his hopeful devotion - his bouyant courage, inspire us to emulate him, however feebly - and whilst weeping for our beloved Chieftain - our sabres must be thrust to the hilt - and our motto be - Independence or Death.

Fitz Lee
Major Gen
Comdg.

While the enemy continued to press on Richmond, Fitz Lee held his command between the Federals and the capital. On May 19, Lee moved from Mechanicsville to Atlee's Station, then learned that the enemy was at Hanover Court House. By the time Lee arrived at that place, only fleeing troopers could be seen after having burned the railroad bridge. The next day, enemy cavalry under Custer returned, only to be driven off once again.

Only five losses of the 1st Stuart Horse during May made it to the muster records. Andrew Connor received a gunshot wound and died at Lynchburg on May 21. John Warro and John N. Moore were captured on May 8. It was Warro's second time; he was exchanged again in November. "Met" Coode and T. A. Slacks, while not injured themselves, lost their horses at Yellow Tavern. "Met" received $1,566.66 in payment for his loss.

Both the Ashby and Lynchburg Beauregard batteries were in action in the next few days while on picket duty on the North Anna River near Pole Cat Creek. However, there is no record of the participation of Johnston's guns in those few days.

On May 25, while resting near Hanover Academy, an inspection was held by John Esten Cooke, as assistant inspector general, of Johnston's and Shoemaker's batteries. In the batteries serving with Fitz Lee, Cooke found unsettling problems with regard to supplies and horses:

> These batteries have lost heavily in killed and wounded horses in the operations against Sheridan, and the horses are in very bad condition. In Johnston's battery 28 have been killed, or wounded so badly as to make it necessary to abandon them; 5 others wounded are with the battery. Of the other horses they are in very bad order, with painful galls on back and neck, and greatly broken down. Two of the gun-carriages have only 5 horses apiece and two others only 4 apiece. Shoemaker's battery is in worse order still; 16 of the horses are covered with sores, and so much broken down by heavy marches as to be unfit for active service. In this battery 4 have been killed, 6 wounded and abandoned, 2 broken down and left on the road, and 2 strayed or stolen. Thirty horses are needed to put this battery in efficient condition. Three mules in ordnance wagons are broken down and unfit for service.

Cooke reported: "The order directed an inspection to ascertain the number of horses unable to "work in harness." "The horses reported are in such an exhausted condition that they are unfit for severe marching or rapid movement, but are still worked in harness." Cooke found that several battery members had been dismounted to supply an ample number of horses for the teams. For the horses available, Cooke opted not to have them turned in, but suggested at the least a few days of rest.

Johnston's battery:
Horses killed, wounded or abandoned: 28
wounded slightly: 5
sore-backed and broken down: 6
Total (cannoneer and gun horses deficient): 39

On May 26, Lee's Division was joined by both Hampton's and Rooney Lee's divisions. Hampton, the senior ranking officer, took command of the cavalry corps on the same day. Many men, including those from the horse artillery, felt that after the loss of Stuart, Hampton could not stir the same emotion in battle as had been customary before.

While the cavalry reorganized with their new commander, Grant continued his advance toward Richmond, through Caroline and King William counties. In turn, Hampton moved toward Hanover Court House.

On May 28, near Haw's Shop, Johnston's and a section of Shoemaker's batteries took a role in the fight that occurred. Due to the terrain, the remaining sections and Thompson's Battery were prevented from participating, though under fire. For nearly two hours the fighting was joined, but by the close of the day, the Confederates had been forced to withdraw to the south of Totopotomoy Creek. The next day, the Confederate cavalry force returned to Atlee's.

Though records do not indicate when or where he was wounded, Thomas Loudenslager suffered the loss of his right arm as a result of amputation on May 30, from a wound.

The following day, Fitz Lee's Division moved via Atlee's Station and Mechanicsville to Cold Harbor. Relieving General M. C. Butler's Brigade on May 31, the men came under attack by Federal cavalry in the afternoon. In the intense heat of June 1, the division was again tried and eventually reinforced by infantry while Grant continued to hammer away at the lines with artillery, cavalry, and infantry. One section of Johnston's Battery was engaged on the same day at Gaines' Mill; another participated in action at Meadow Bridge. The cavalry and horse artillery were finally relieved on June 2 with the arrival of General John C. Breckinridge's infantry division. Charles Branch was the only man recorded in the rolls as wounded on June 2. Branch suffered a shell wound to his arm and his horse had been killed as well.

For the next several days, Johnston's Battery bivouacked and stood picket duty, along with the rest of the division, on the Chickahominy. On June 8, word was received that Sheridan had crossed the Pamunkey and was moving into King William County. Leaving Rooney Lee to guard the Chickahominy, Hampton moved at once with his old division and that of Fitz

Lee to intercept the Federals. Riding steadily, Hampton slipped between Sheridan and his suspected targets of Gordonsville and Charlottesville. Moving along the Brook Turnpike, the group changed course along the Telegraph Road at Yellow Tavern toward Louisa Court House. After halting briefly at Hewett's to feed the horses, the Confederates marched in the dark to Beaver Dam Station and made camp at midnight.

The following morning the march resumed, passing Frederick Hall and Louisa Court House. By midnight the contingent had reached a point five miles west of Trevilian's on the Charlottesville Road, where the men and horses rested for the night.

By the time the batteries arrived at Trevilian's on the morning of June 11, skirmishers had already opened along the line. Rosser's Brigade moved in the direction of Gordonsville to guard the left flank, but back toward the center of the battle the Federal cavalry managed to breach the Confederate line and take two caissons and several horses of dismounted men. In an effort to halt the Federal momentum, General Rosser charged with his brigade into the fray and dispersed the enemy, reclaiming the previously lost material, horses, and men. Johnston's Battery had one section engaged along with a section of Shoemaker's.

On June 12, Sheridan renewed the attack. A section of Johnston's and one gun from Shoemaker's Battery participated against the enemy in the afternoon and aided in turning the enemy's right. Captain Shoemaker recollected that the position was in "some dense shrubbery on the enemy's right flank..." Once the sections had opened, Fitz Lee rushed his cavalry headlong, throwing the Federals back. Shoemaker, once again, recalled that it was at this point that the batteries "returned" the ammunition captured from the enemy on the previous day.

On the following day, the Confederate cavalry started in pursuit of the enemy raiders. Moving down the Central Railroad, through Louisa Court House and back toward Frederick Hall, the pursuit continued across Cushionberry's Ford on the North Anna on June 14 and past Chilesburg the next day. "Met" Coode suffered his second wound on June 15 while on the move. Upon arriving at White House on June 20, eight guns from Thomson's, Hart's, Shoemaker's, and Johnston's (one section) batteries were engaged for the entire day against enemy fortifications. A masked Federal battery announced its presence and gave the Confederate gunners return fire, soon coupled with that of two gunboats which made their way up to the scene. After about an hour, competing with the gunboats and one of their 32-pounders seemed fruitless, convincing the horse artillerymen to conserve their strength and ammunition.

Near nightfall, the battalion relocated to a point one mile down-river and camped for the night. The following day, the enemy attempted an attack upon the camp, but were briefly and quickly stopped by a gun from Shoemaker's Battery. The lone gun signaled the cavalry and gave them time to assemble. The bulk of the battalion withdrew two miles and re-formed. At St. Peter's Church the cavalry and guns went into line and soon encountered a large blue wave of the enemy. Although the Confederates fired for about an hour, the Federals showed no sign of wavering, forcing the Confederate contingent to retire during the afternoon.

On June 21 the battalion was again divided when a large enemy cavalry force under Generals J. H. Wilson and A. V. Kautz struck the Petersburg and Weldon Railroad at Reams' Station. General W. H. F. Lee immediately led his division, including Shoemaker's and Johnston's batteries, on a hot and dusty pursuit. On the morning of the 22nd the enemy passed by the Southern right and moved in the rear before they could be intercepted. The enemy then continued to move along the line of the railroad toward Danville. On June 23 the division caught up with the enemy near Black's and White's (now Blackstone). Johnston's Battery and one gun of Shoemaker's engaged the enemy near Forge Bridge, a short distance from Nance's Shop. The engagement was again joined the next day near Nance's Shop and the Confederates successfully routed the enemy. Major Breathed was seriously wounded on June 26, prompting a note from General Robert E. Lee hoping for his recovery soon. Breathed would be back in the field by the close of the summer.

On June 26 camp was made near Chaffin's Bluff and on the 28th the batteries crossed the James River and camped near Petersburg. The following day the contingent was in camp at Reams' Station, in which place it would remain until the end of July, primarily performing picket duty. In their position near Reams' Station and Stony Creek Station, Fitz Lee's cavalry division protected the extreme right of Lee's badly overextended army.

On July 30 the battalion was roused by the great explosion of the Crater, but remained in camp and only heard of the fighting that took place. The following day the horse artillery went in pursuit of enemy cavalry that was bearing down on Reams' Station. By the time the horse artillery arrived the threat had disappeared. Camp was resumed that evening near Reams' Station, where the battalion again was visited by John Esten Cooke. To the delight of the inspector-general, the battalion assembled its glee club in the afternoon and serenaded the honored guest with its best renditions.

On August 5, Fitz Lee's division and Kershaw's infantry division, along with Johnston's and Shoemaker's batteries, marched under orders to report to General Jubal Early in the Shenandoah Valley. The horse artillery's guns were loaded on cars of the Orange and Alexandria Railroad for trans-

Map, Action at Guard Hill, August 16, 1864

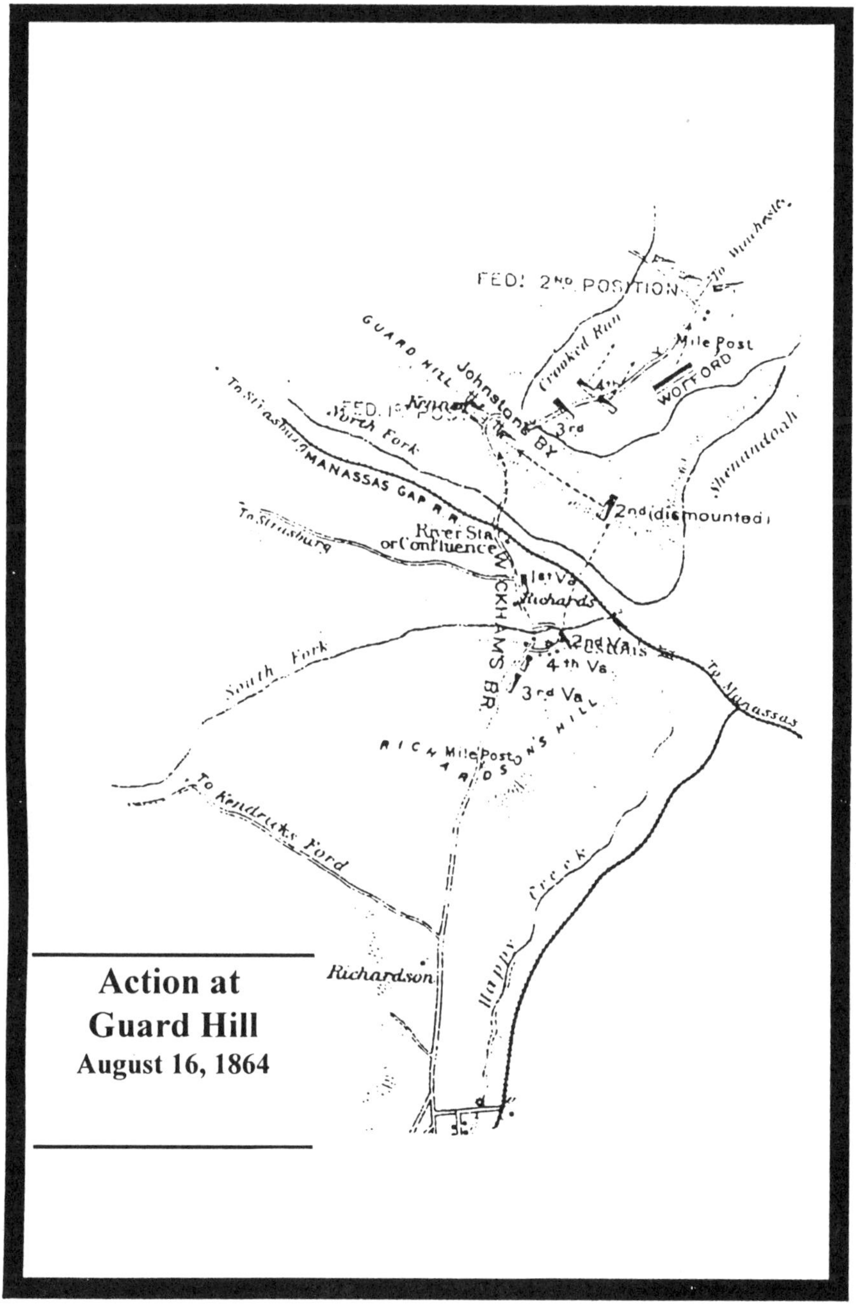

Action at Guard Hill
August 16, 1864

Map, Action at Spout Spring, August 19, 1864

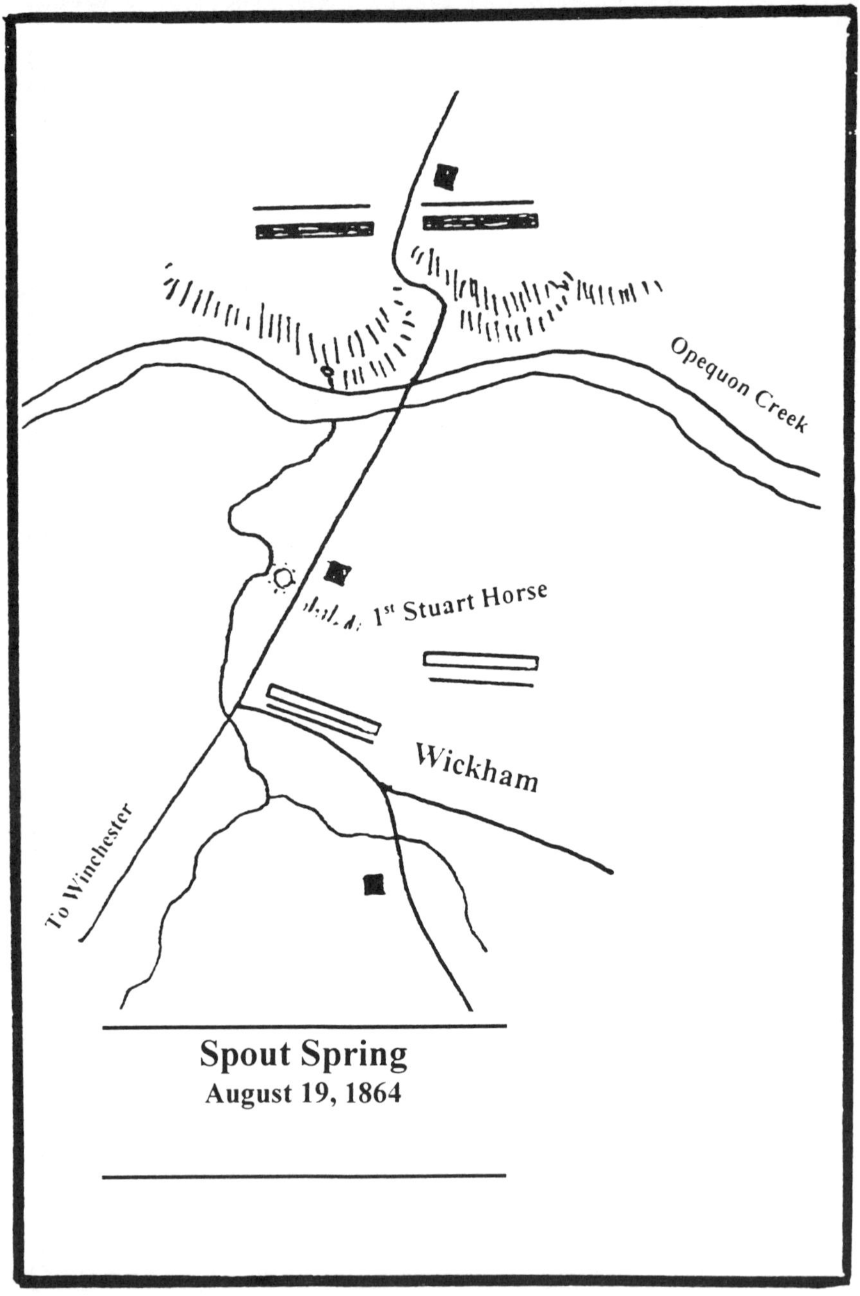

portation to Mitchell's Station. The men reached Culpeper Court House by horse on the 11th and reclaimed their pieces. Leaving Culpeper on the 12th, the batteries arrived at Front Royal on August 14, where they camped for three days before moving on to the next bivouac at Winchester on August 18. Still recovering from his wound, Breathed was temporarily succeeded in command of the battalion by Captain Johnston. In Johnston's place, Lieutenant Daniel Shanks became interim commander of the 1st Stuart Horse Artillery.

After but little time for a respite, the horse artillery battalion was soon on the march again in the Shenandoah Valley. Like the infamous "Ten Days" of November, 1862, the 1st Stuart Horse Artillery was in action nearly daily. Fighting for the battery appears to have began on August 19 at Spout Spring, driving the enemy back with the Wickham's Brigade past Berryville. George W. Covington was the only man recorded as having been wounded at Berryville, but on August 21. Covington died as a result of his wounds on September 13.

Marching via Charlestown, Leetown, and Martinsburg to Hainesville, the 1st Stuart Horse accompanied its brigade into action again on August 25 at Williamsport. After running the enemy cavalry across the river, one member of the 1st Virginia Cavalry recollected "Our Artillery shelled Williamsport...."

By September 1, Johnston's Battery had apparently left the Valley temporarily and was recorded as being at Gordonsville, perhaps to replenish the stock of horses. By the middle of September, the 1st Stuart Horse Artillery had returned to the Shenandoah Valley.

On September 19, Sheridan attacked Early on the Berryville road. In what erupted into the Third Battle of Winchester, Johnston's batteries were heavily engaged. The 1st Stuart Horse Artillery was posted to the left with Fitz Lee across Red Bud Run. In time, however, the full force of Sheridan's army came crashing down on the Confederates. Moving across the battlefield from point to point where needed, Johnston's Battery was reported to be "operating with good effect." The only loss reported in the muster records was that of H. R. Smith's horse valued at $2,333.33.

By the close of the day, Early was forced to retire from the field. Acting as the rearguard, the cavalry kept the enemy at bay until reaching Middletown. From there, Fitz Lee's Division moved up the Luray Valley by way of Front Royal. On September 20, Sheridan sent A. T. A. Torbert's Third Division under Wilson to intercept the Confederate retreat from Fisher's Hill. Johnston's Battery, continuing to accompany Fitz Lee's Division (then under Wickham), first encountered the enemy that day at Riverton on the Shenandoah River. Though the enemy batteries shelled the Confeder-

Map, Engagement at Williamsport, August 25, 1864

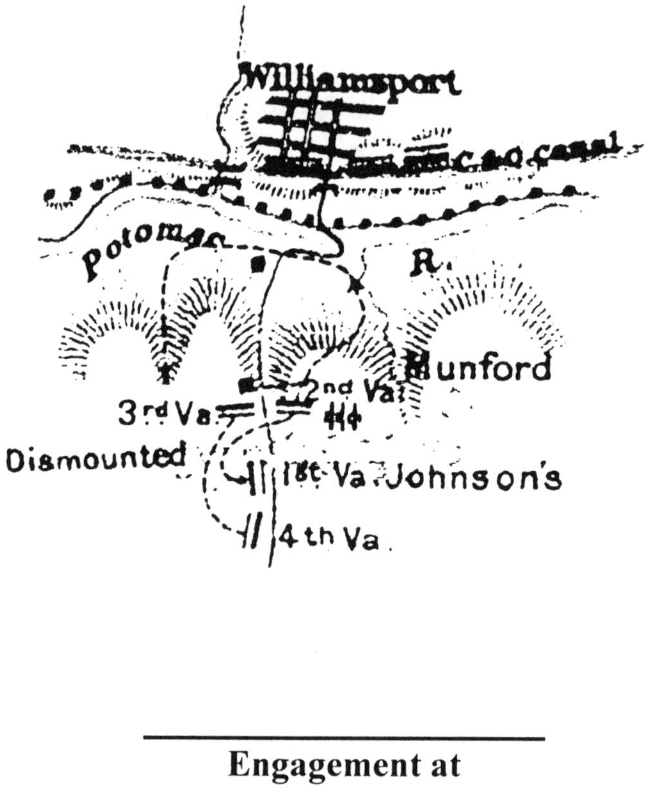

Engagement at Williamsport, VA. & MD.

August 25, 1864

Map, The Shenandoah Valley, 1864

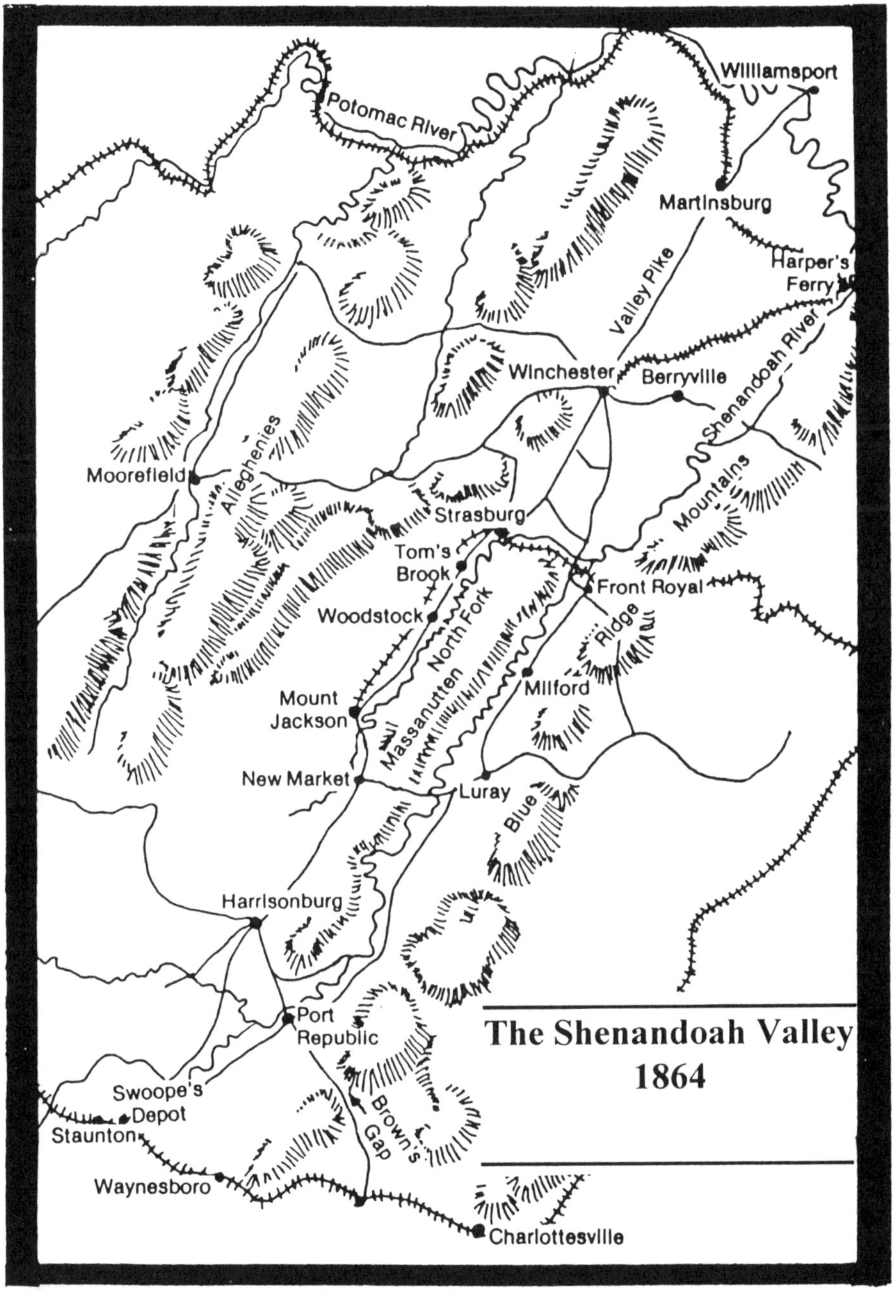

Map, Winchester, September 19, 1864

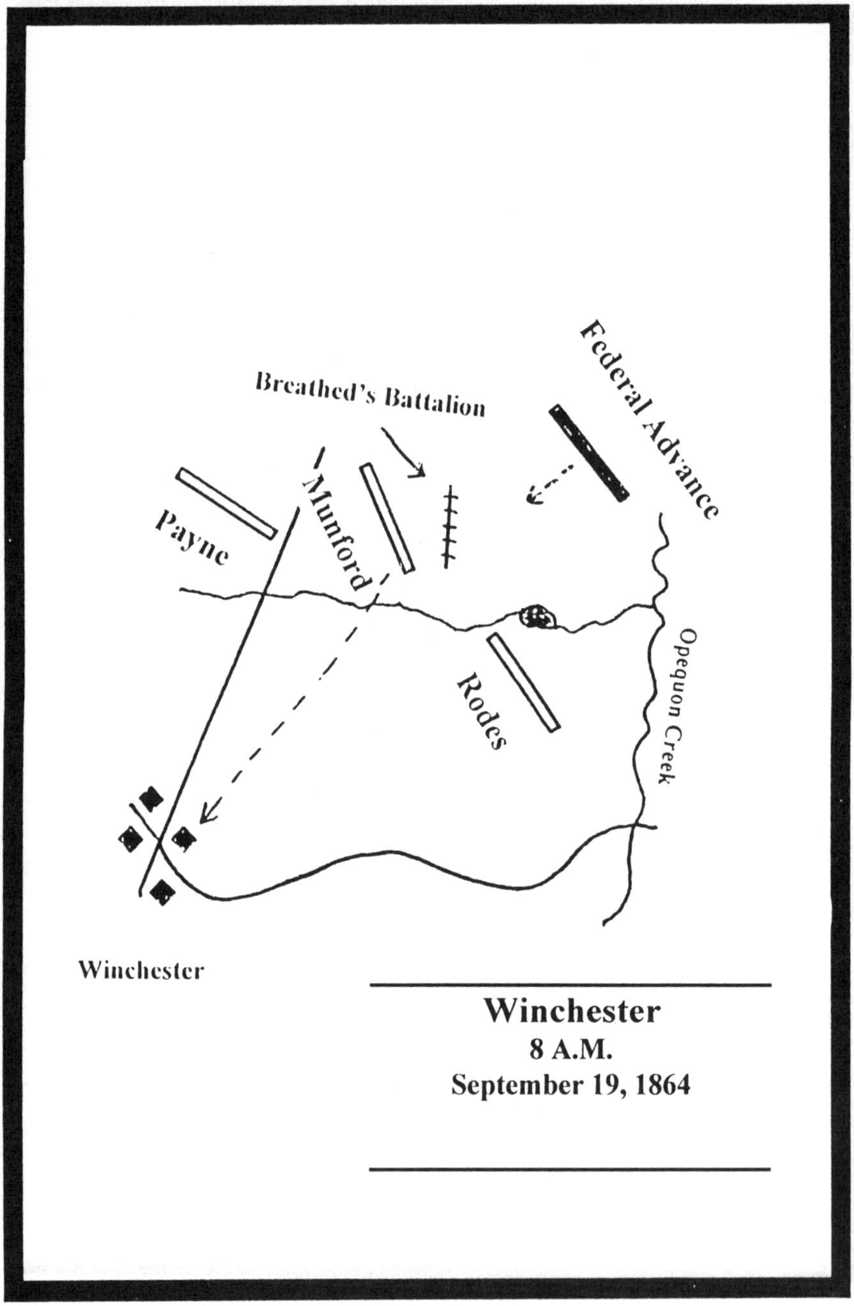

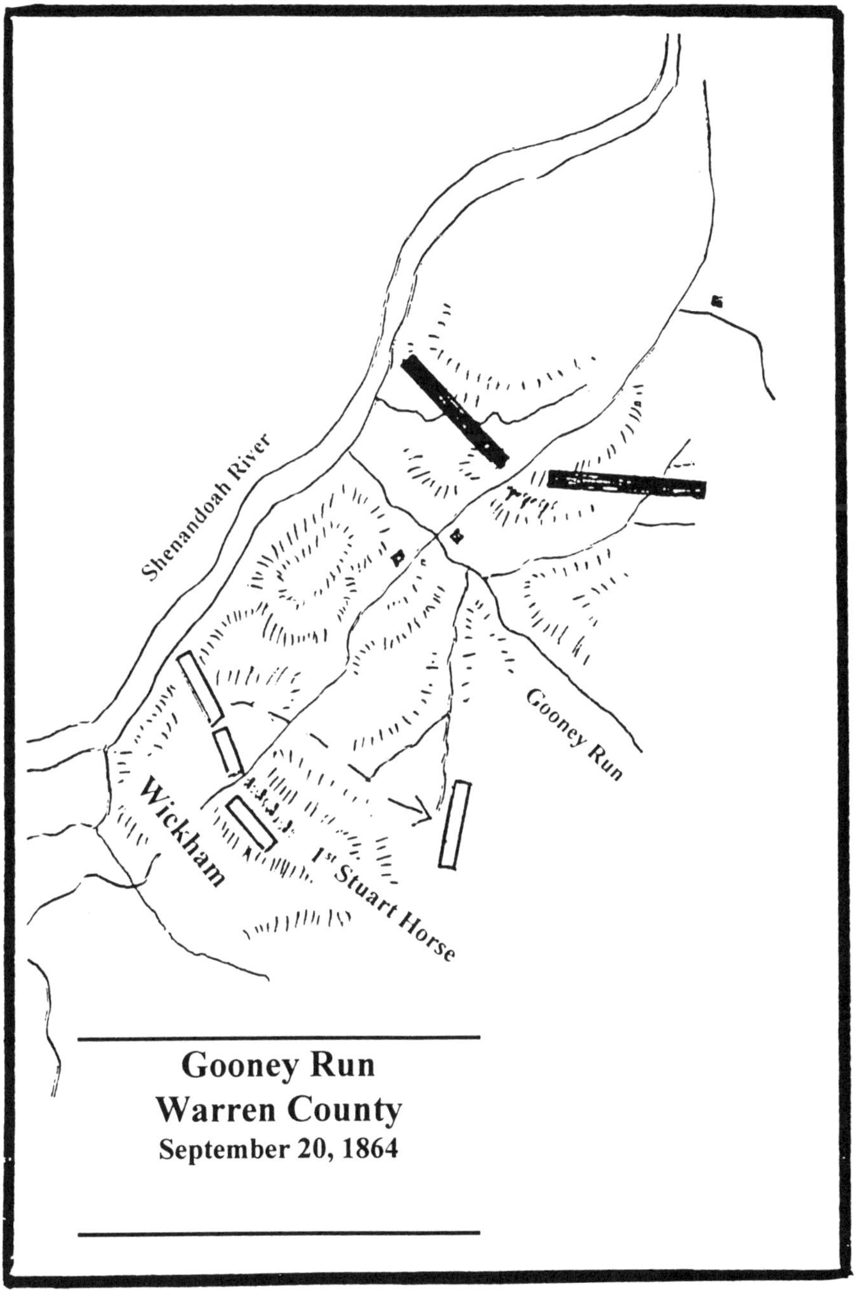
Shenandoah River
Gooney Run
Wickham
1st Stuart Horse
Gooney Run
Warren County
September 20, 1864

Map, Milford, September 21, 1864

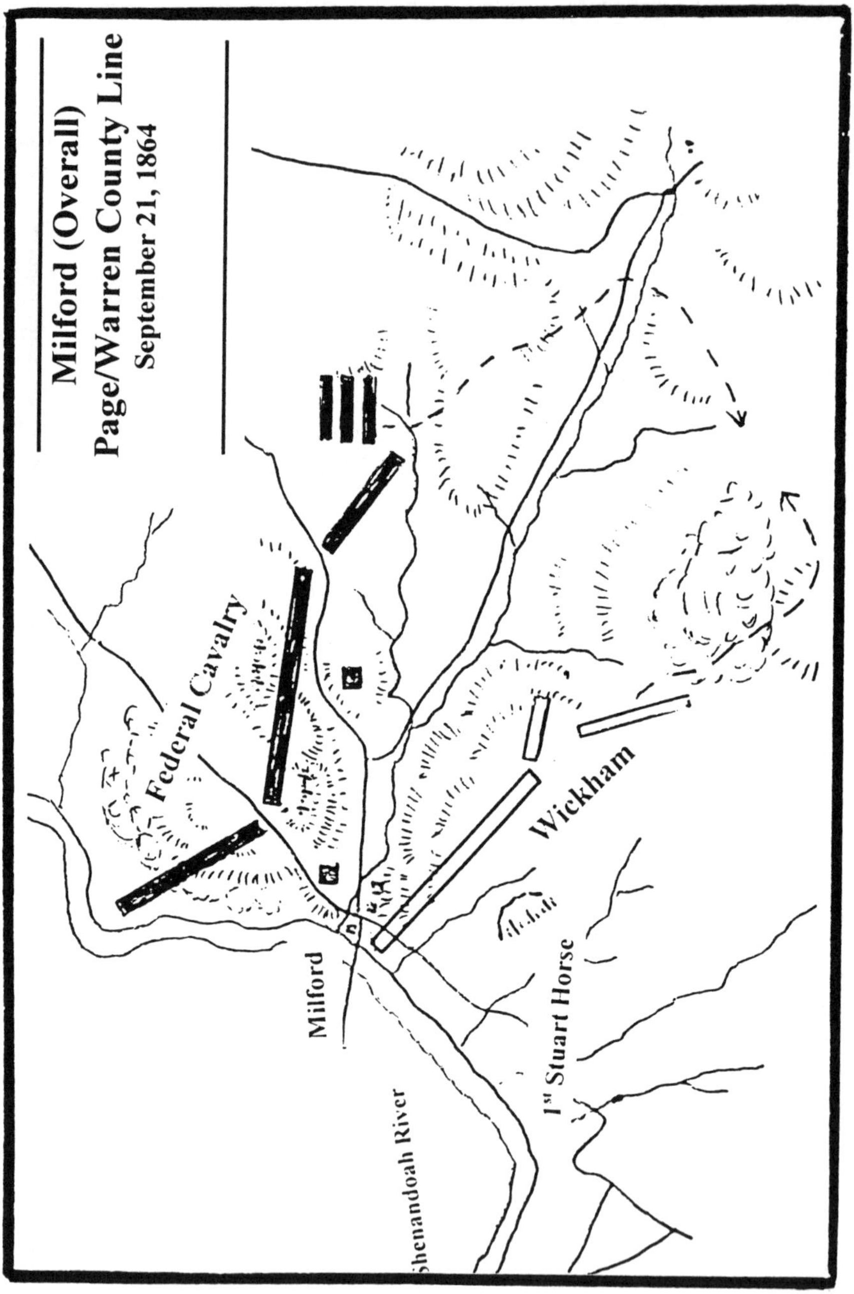

ates, the distance was too great for a general duel of artillery. The horse battery, however, apparently did not engage the enemy until the following late afternoon. Captain Johnston's horse was listed as killed on September 20.

On September 21, the Federal cavalry again moved against Wickham's Brigade. In the difficult terrain along the Warren and Page County line, little result would come of the awkward affair. As darkness settled in on the hills around Milford (now Overall), Johnston's battery was sent up to give an opposing battery counter-fire from which a duel commenced for nearly one hour.

That evening the Confederates dug in near the little village on the county line while another division of cavalry under Wesley Merritt moved to reinforce Wilson. By 2 a.m. on September 22, the Federal force had merged and moved up the Luray Pike to strike Wickham. By then the Confederates had withdrawn to better ground on the south side of Milford Creek. The position would prove impossible to attack directly or practicably by any other method. Johnston's Battery continued to participate in the fight during the day with no losses. That evening the Federal cavalry withdrew from the field.

On September 23 Wickham pulled away from Milford. The Federals, receiving news of the victory at Fisher's Hill immediately set out to strike Wickham again and finally caught up with his contingent near Luray. After a charge of the enemy's First Brigade, led by George Custer, the Confederates gave up the position and moved toward New Market. By dusk the Federals, having driven the Confederates and captured prisoners and a stand of colors from the 6th Virginia, withdrew from the valley by way of a mountain road.

Finally, on September 27, Wickham marched to rejoin the rest of Early's army; but finding enemy cavalry moving through Massanutten Gap, he was forced to fall back into Luray Valley. Opting for an alternate route, Wickham's gray troopers reached Port Republic the next day. Withdrawing his army into Brown's Gap, Early set out picket lines and temporarily recovered from the thrashing by Sheridan. Worn out from the rapid movements in the Valley, most of the horse artillery retired on September 28 to Greenwood Tunnel, where they remained until early October securing fresh horses. By October 4, Johnston's battery was again on the move.

Rapidly reacting to the atrocity of Sheridan's burnings, the 1st Stuart Horse remained with Wickham's Brigade engaging the enemy on October 4 on the North River. Unable to strike a fatal blow against Early, Sheridan began retiring down the Valley again on October 5. As Sheridan moved away, Confederate cavalry struck the enemy rear, moving by way of the

Action Weyer's Cave, September 27, 1864

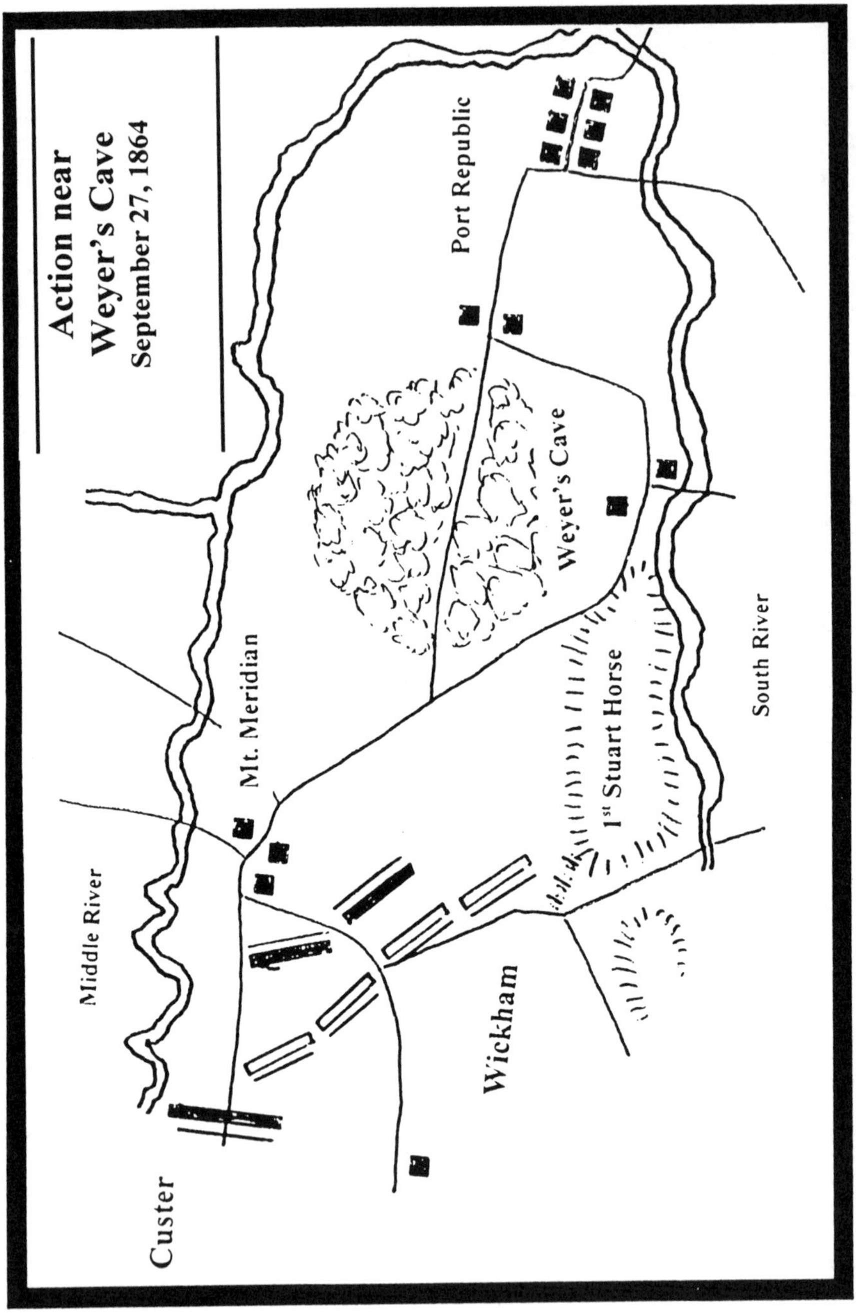

Action at Waynesboro, September 28, 1864

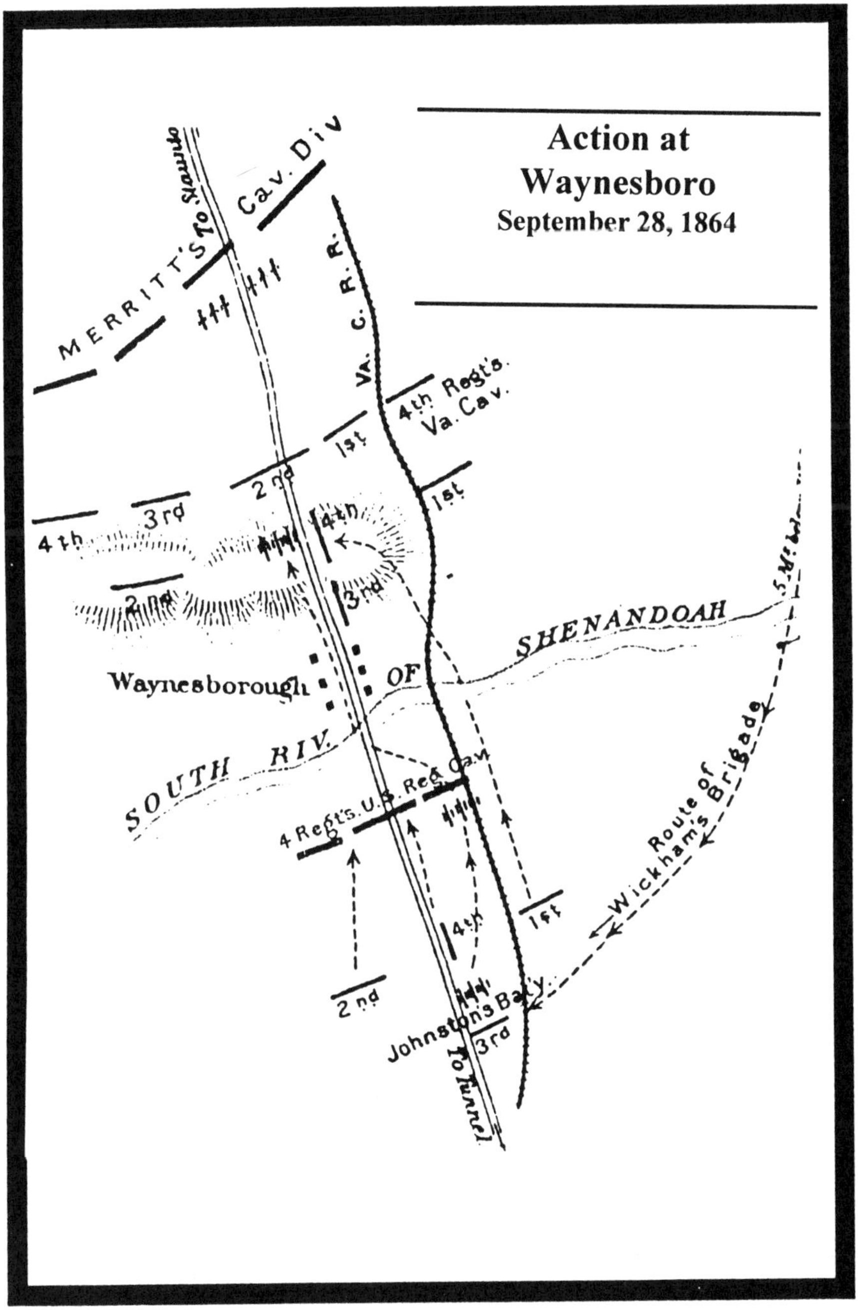

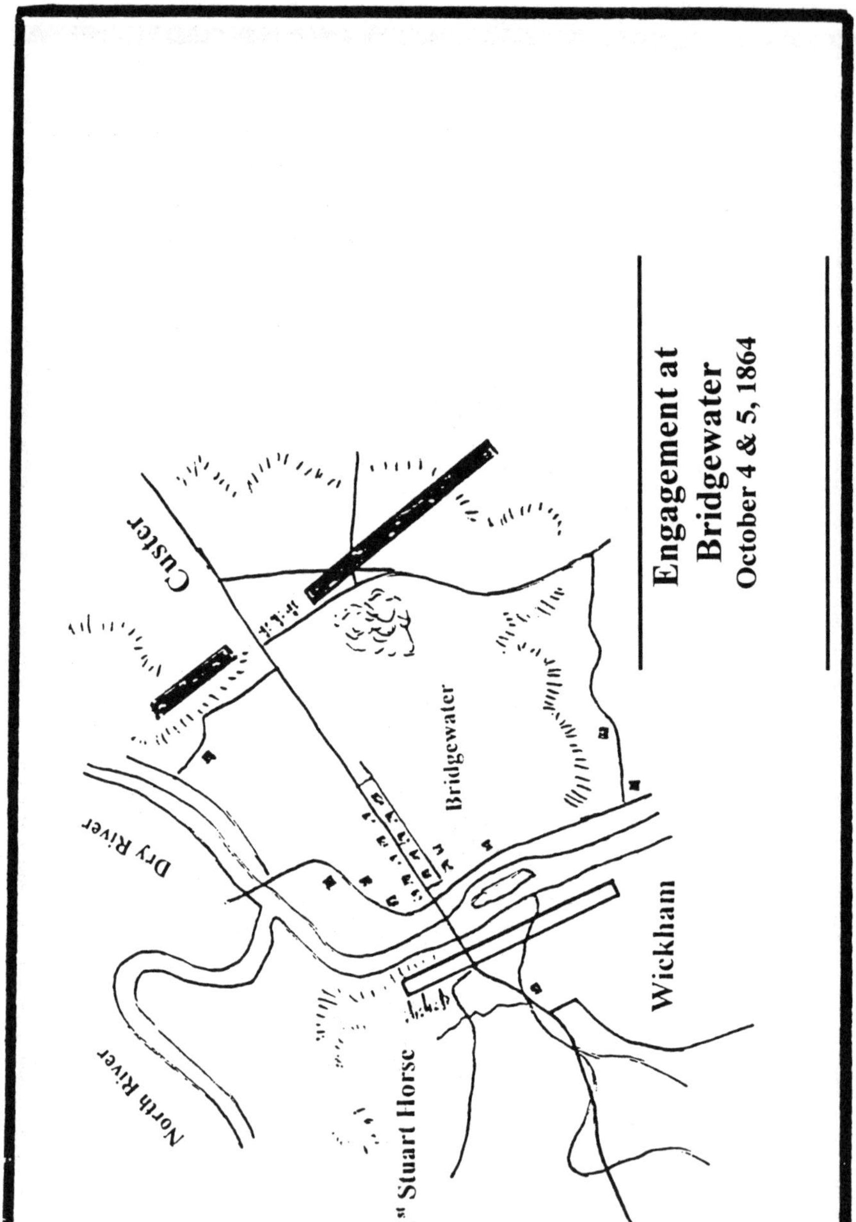
Engagement at
Bridgewater
October 4 & 5, 1864
Custer
Bridgewater
Dry River
North River
Wickham
1st Stuart Horse

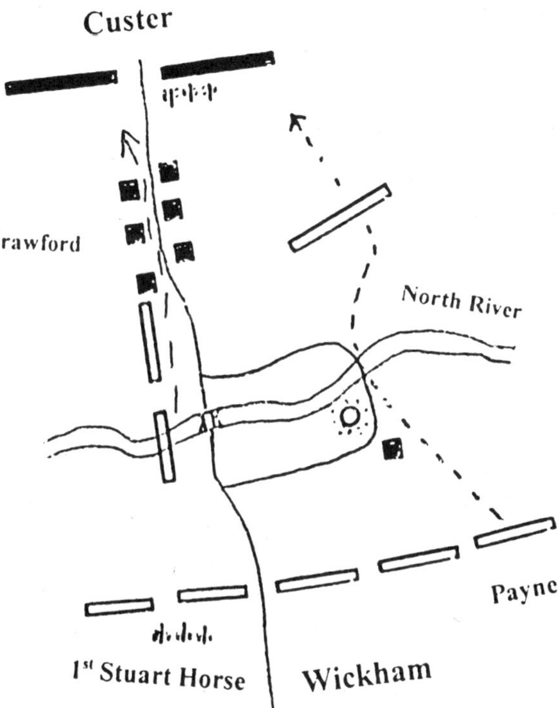

Engagement near Mt. Crawford
October 4 & 5, 1864

Back Road to Brock's Gap. The battery was again engaged on October 5 near Bridgewater. Private P. H. Keleher was recorded as wounded in the right thigh in the movement on October 4.

Confederate cavalry set out in pursuit under a smoke filled sky. Sheridan made certain that his wake was full of destruction. On October 9, the cavalry again clashed with the enemy at Tom's Brook. Placed on a high hill in the rear of the cavalry, the horse artillery stood without any support. The Ashby and 1st Stuart Horse Artillery joined in the fight. However, the Confederate skirmish line could not bear the brunt of the onslaught of Federal cavalry, severely threatening the Confederate horse artillery. The guns were reported to have been used "with good effect to the last." Although many gunners had been able to cut their way out of the harrowing situation, six guns were lost.

Following the fight, the 1st Stuart Horse Artillery hardly had enough men and equipment to field two guns. At least 23 were captured, including Lloyd Beall, William V. Bennett, N. Coit, Robert Crabb, William Elam, N. C. Glenn, W. A. Grubbs, Alex Harris, William Harrison, Hayden Haynes (died at Pt. Lookout, 5/8/65), Virgil Haynes (died at Pt. Lookout, 2/14/65), James Higgins (Higgins' horse had also been killed), William Lewis, Elijah Lindsey, James S. Lindsey, W. W. Mangum, R. J. Roach, B. F. Miller, Nicholas Morton, James Ryan, William G. Smith, J. Supinger, W. C. Watkins. All were conveyed to Harper's Ferry and then to Point Lookout, where many remained to the close of the war. A few, however, were fortunate enough to be exchanged in March 1865. Though not captured, Leonidas Smith suffered a wound that would cause his death eight days later. Dennis Conroy suffered the loss of his bay horse that had been valued highest in the company at $3,000.

The following day, Shoemaker's Battery and the remaining section of Thomson's accompanied Lunsford Lomax's cavalry to guard a wagon train on the Valley Pike near Woodstock. The ensuing action nearly cost all of the battalion's guns. Had it not been for the "extraordinary gallantry of the artillery officers and men," all but one of Thomson's guns might have been lost. Captain Carpenter of Braxton's battalion was particularly noted in after-action reports for his efforts in saving the guns. For his efforts, Carpenter received a wound which necessitated amputation of his arm.

At this point, Breathed's horse artillery battalion had been severely depleted by losses of men, horses, and guns. Nevertheless, the balance of the batteries were utilized to the fullest. For the duration of the month, what was left of the 1st Stuart Horse Artillery moved back down the Valley to Fisher's Hill once again. On October 19, the battalion accompanied Early in his last great effort against Sheridan's army. Though the Confederates

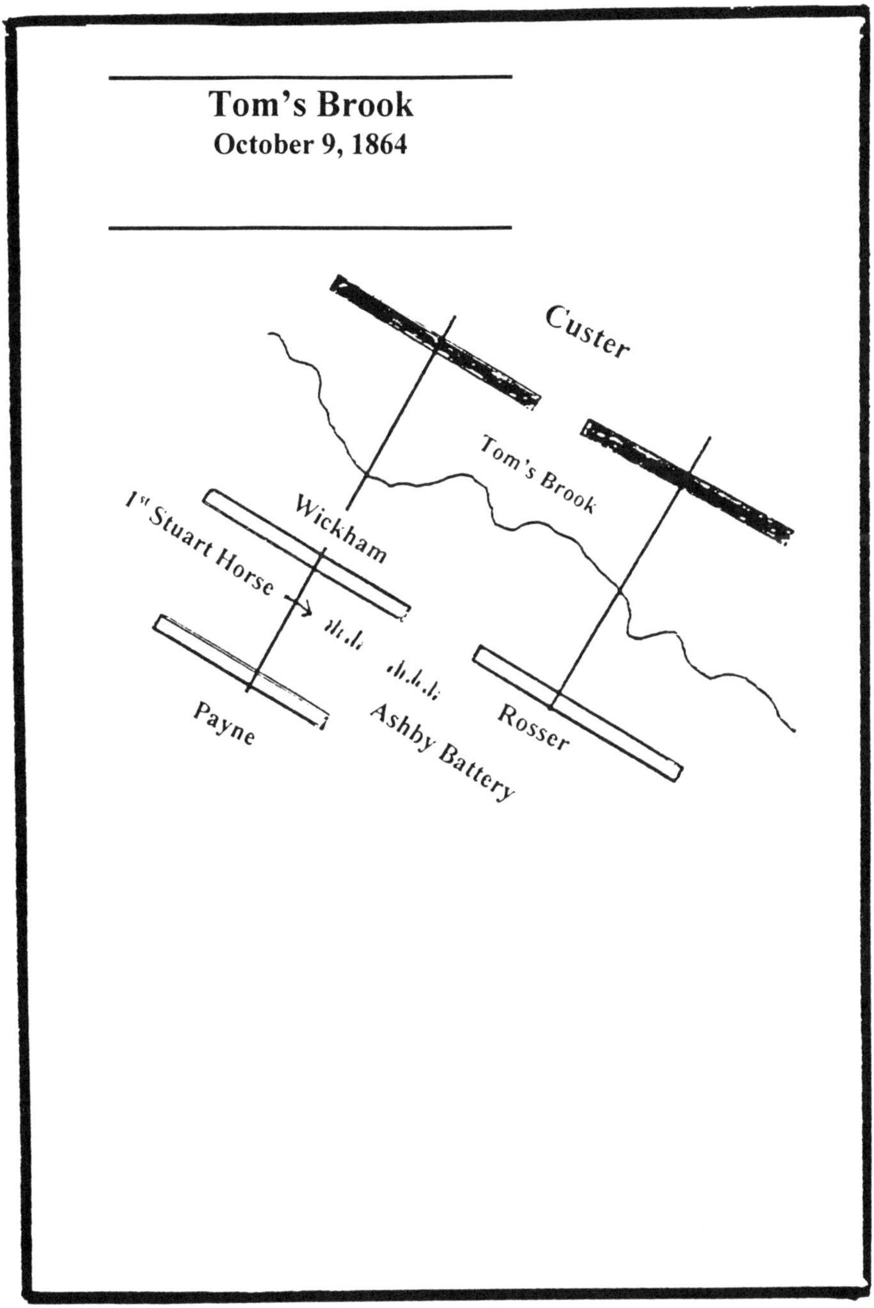
Tom's Brook
October 9, 1864
Custer
Tom's Brook
Wickham
1st Stuart Horse
Payne
Ashby Battery
Rosser

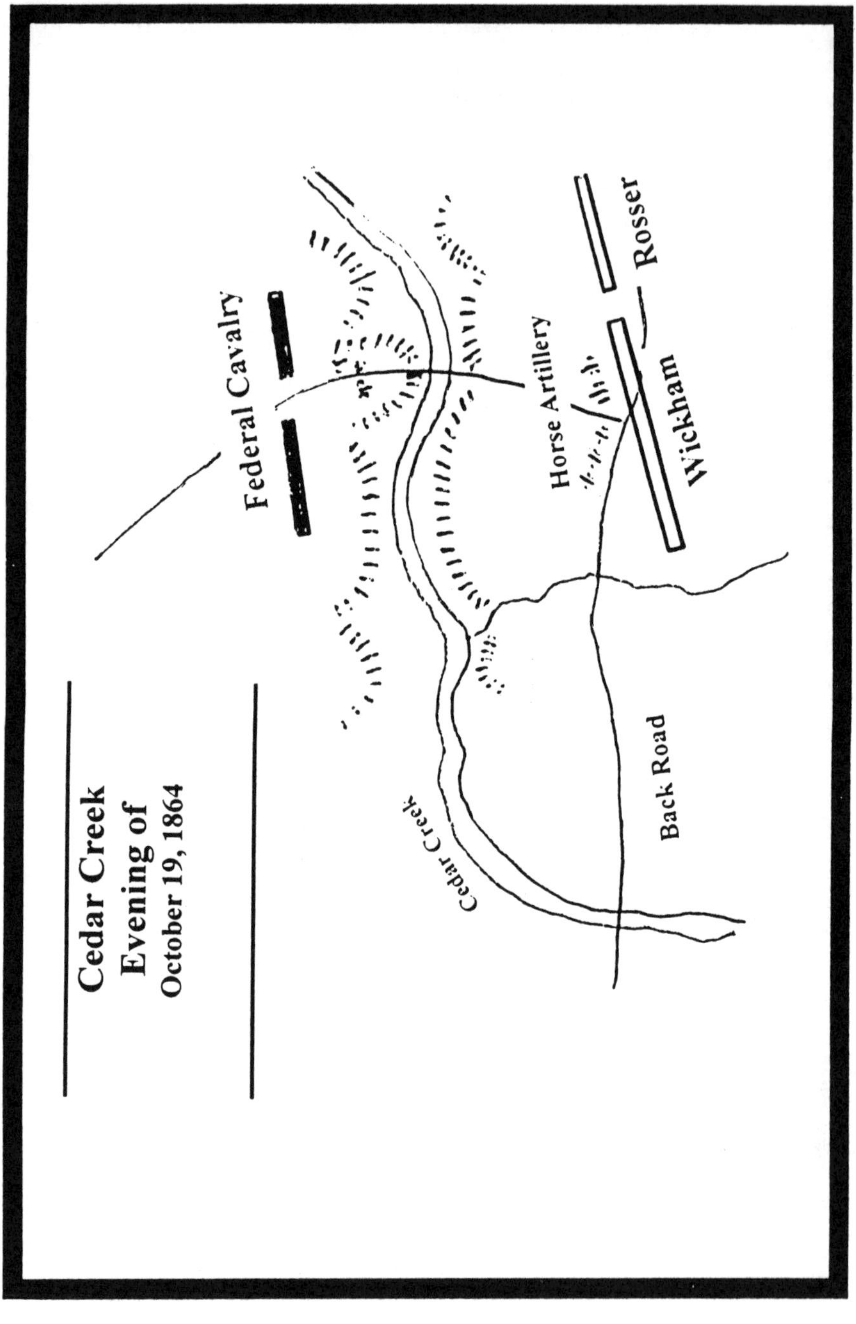
Cedar Creek
Evening of
October 19, 1864
Federal Cavalry
Rosser
Horse Artillery
Wickham
Back Road
Cedar Creek

enjoyed success in the morning, the Federals found their courage late in the afternoon. Again, the battalion nearly lost more guns in the ensuing retreat.

On October 22, the battalion moved to Forestville and camped for several days. What sections were left were sent out on various picket duties at Columbia Furnace and throughout the central Valley. On September 29, Captain William Lyman of the battalion brought in ten dismounted cavalrymen "to fill out the battalion," the numbers having reached such a low point.

On November 9, Johnston's and Thomson's batteries arrived in the Page Valley to join Shoemaker's Battery. The battalion now consisted of only eight guns, including the two guns remaining in Johnston's Battery. That night the battalion received orders to move the following day and to cook three-days' rations. The next morning the battalion moved through Mt. Jackson and Woodstock and camped at Tom's Brook in the evening. Early once again was moving forward with the remnants of his army to confront Sheridan.

At 4 p.m. on November 12 the battalion rode toward Newtown, where the men stopped at dark within a short distance of the enemy, only to move back across Cedar Creek. The following day, the movement resumed and the battalion joined the main army near Carson's Farm. Unable to draw Sheridan out for a fight, Early again retired back up the Valley.

On November 14, the battalion was back in its old camp at Mt. Jackson. Private Charles McVicar of the Ashby Battery wrote in his diary that day that Major Breathed had issued orders for a daily regimen:

> Our daily routine is roll call at daylight. Stable call at 6 o'c, guard mounting for field and battalion guard at 8 o'c, inspection at 9 o'c, sick call at 9 ½ o'c, drill at 10 o'c, and 2 o'c, roll call at noon, catch up horses at 4 o'c, roll again at 6 o'c. Three roll calls daily and this is a reserve recruiting camp.

On November 16 Breathed's battalion was ordered to Mt. Solon and went into camp within three miles of Harrisonburg. With winter closing in rapidly, the frequency of scouting missions decreased a great deal. Snow and rain ruled much of the month of November. By the last of the month Breathed's battalion had moved to camp at Anderson's Farm on the Middle River, six miles from Staunton. Desertions had become a problem since soon after their arrival near Staunton. At least two men, Henry Montenegro and Harper's Ferry native Edward Straney, deserted. Montenegro went on to join the 2nd New York Cavalry.

On December 9 the battalion moved with the Laurel Brigade to P. Koyner's farm on the South River, six miles from Waynesboro on the Port Republic road. The evening of the next day the men scrambled to fix some winter quarters with heavy snow bearing down upon them. Rations were also a scarce commodity during this time. On December 13, McVicar wrote: "Out of rations, bought a half bushel of shorts and midlings mixed for bread."

On December 18 camp was moved again to the H. Teiford farm, seven miles west of Staunton near Swoope's Depot, to build official winter quarters. Snow had been falling on and off for days and covered the ground. McVicar wrote that trees were immediately felled.

In response to an alarm that Custer was moving in the Valley, several sections of the battalion were deployed with the cavalry toward Harrisonburg, only to return shortly afterward. On December 23 the men rushed out of their camp with the Laurel Brigade through Staunton and along the Lexington road to Charlottesville. The next day the battalion returned to its winter home and resumed building. On Christmas morning several men in the batteries enjoyed warm huts and reflected on the days of winter quarters past while consuming poor rations of hardtack and pork.

There was little to celebrate this Christmas other than being alive.

Chapter 7
June - December, 1864
The 2nd Stuart Horse Artillery

Before dawn on May 12, as the Confederate line around Spotsylvania Court House had been breached by an incredible assault, McGregor's Battery stood poised with a picket from Chambliss' cavalry on the Myers Farm. Since the unit had served as pickets for the past two days, it appeared that the week might pass with little incident. However, on the night of the 12th, McGregor's Battery along with Chambliss' entire brigade moved out toward Fredericksburg on the Telegraph Road. The following day the battery engaged on and off, losing three men killed. At least two men, John A. Donald and James King, were listed on muster records as killed on May 12. Reuben Nuckols may have also been wounded during the action. There is a confusion in the records as to whether it was on May 9 or May 12.

Sometime during the fighting around Spotsylvania, Private Cardwell of McGregor's Battery was on leave and heard of the action. At once Cardwell set out on his return to his company and after an arduous journey arrived on the field as his battery was engaged. "As luck would have it," Cardwell ran right into the action where his battery was busy at work. He heard all sorts of surprised comments from his comrades as to where he had come from so suddenly.

On the morning of May 14, the battery was again at Myers Farm as a picket. With little more support available, the battery's position was shortly tested by the 140th New York and 91st Pennsylvania of Ayres' Brigade. However, the first trial would be the batteries of Cooper and Breck, numbering twelve guns. McGregor's men quickly manned their pieces and fired on the enemy guns opposite them on the Beverley homestead. McGregor's men soon sighted the oncoming infantry and immediately sent a courier to Chambliss' full brigade stationed about a mile to the southwest. Upon the arrival of the cavalry, the men dismounted and offered gallant but futile resistance against such a large number of Federal infantry. As musketry and cannon fire increased, the Confederates were forced to withdraw from the farm.

Sometime after 4 p.m., Confederate infantry under A. R. Wright moved against the defending Federal force on the Myers Farm. The previous enemy contingent had been relieved, leaving Myer's Farm now defended by infantry under Emory Upton, behind recently thrown up breastworks. As the Confederate infantry advanced, Chambliss' force moved opposite the left of Upton's line. Soon after, the infantry charged the dug-in Federals and McGregor's guns revealed themselves to deliver a severe fire on the flank. The hill was soon, once again, in Confederate hands.

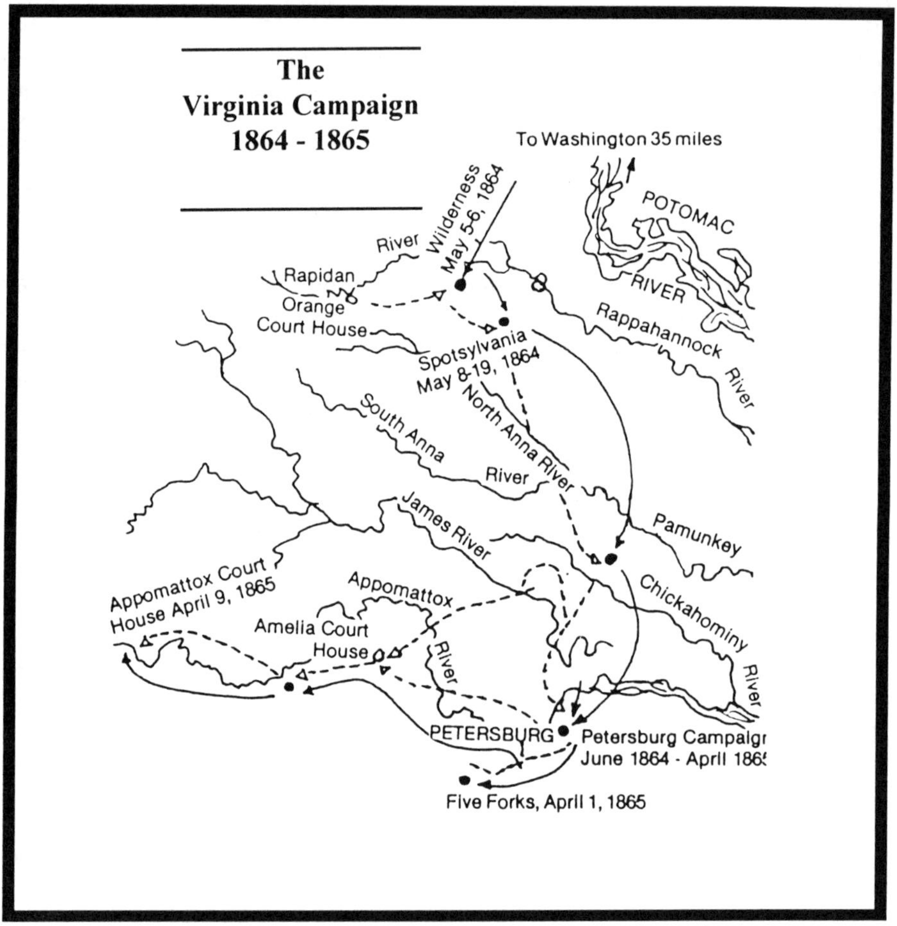
The
Virginia Campaign
1864 - 1865
To Washington 35 miles
POTOMAC
RIVER
Wilderness
May 5-6, 1864
River
Rapidan
Orange
Court House
Rappahannock
River
Spotsylvania
May 8-19, 1864
North Anna River
South Anna
River
Pamunkey
James River
Chickahominy
River
Appomattox Court
House April 9, 1865
Appomattox
River
Amelia Court
House
PETERSBURG
Petersburg Campaign
June 1864 - April 1865
Five Forks, April 1, 1865

While Grant attempted to flank Lee, Chambliss' Brigade and the 2nd Stuart Horse Artillery maneuvered against the enemy column, riding down the Telegraph Road past Chilesburg to Stanard's Mill. Battalion Ordnance Sergeant Walter Dabney was captured during the march on May 15. Though the enemy drove in the cavalry pickets, the bulk of the cavalry and McGregor's Battery held in position. The enemy appeared but withdrew within a half hour after the battery had opened. Shreve commented on the action of the 17th: "Late this evening in same position, but the enemy fell back after a few shots." From the 17th until the 21st the Federals made half-hearted attacks on the position of the battery but were pushed back each time with little effort. Shreve again commented of the 18th, "We were attacked this evening in same position, but the enemy fell back after a few shots." On the 21st, the battery had "orders to be ready to fire at any minute to-night."

The next day, McGregor's Battery, still accompanying Chambliss, was engaged near Mud Tavern and at Ashland the next day. Striking a corps of Federal infantry, a member of Chambliss' Brigade wrote: "The Tenth Regiment in front dismounted and charged the woods, which skirted the Telegraph road. As the squadrons of the ninth were formed in close order, the men of the Tenth were pressed back, fighting. Two of our squadrons here hastily dismounted and [were] thrown forward to the right. The rest, with drawn sabres, were held ready to charge. Our guns now opened fire, and more rapid discharges we never heard." Shreve boasted that they "made some good shots at the Federal Infantry, while they were passing to our right." By nightfall the batteries were near Beaver Dam Station.

Marching and counter-marching on May 23, the battery finally settled in with the brigade for several days along the North Anna River.

On May 27 the battery was again on the move with Chambliss, through Ashland and out Old Church road, where they camped. The next day at Haw's Shop, Fitz Lee and Wade Hampton dismounted their troopers and placed them to halt Grant's advance. After engaging the enemy, that evening the Confederates were again obliged to withdraw. Shreve reflected that evening on the sadness felt over Stuart's death: "we feel depressed at the loss of General Stuart, while Hampton is a good fighter, we have not that confidence in him that we had in Stuart."

On May 30 the enemy continued the right flank maneuver. The 2nd Stuart Horse Artillery followed with a squadron of cavalry, and fired one gun against them as they moved. While Lee continued to retreat, McGregor's Battery would continually be involved in the "shoot and retire: routine.

On the afternoon of May 31, McGregor's Battery was placed into position in a "large level field" just south of Hanover Court House. While quietly holding the position, the battery received orders to limber up and move into Richmond. Shortly after the march began, a courier caught up with the company and told them they had to return at once to their previous position. Turning around, the battery moved back at a trot, but soon was ordered in at a gallop. Upon re-arriving at the field, the place was being "swept with rifle bullets." Private Cardwell:

> We dragged two dead horses into action and went into action.... and began firing short-fuse shells and in a short time changed to canister.... As I recall the 10th Virginia Cavalry galloped up as our support, and Lieutenant Bob Caskey's [Caskie] horse, a fine gray, went down with the gallant Colonel under him. We did not stop to help; we were too busy.
>
> The enemy charged us several times and was repulsed. The firing was continuous and hot. Captain McGregor was on his horse near a small clump of bushes to the right of the battery a little in front and he was making a good deal of noise shouting: "Give 'em hell, boys! Pour it in, boys!"

Lieutenant Ford meanwhile rode calmly behind the guns supervising the fire. Darkness was quickly setting in and with each flash of the guns, Cardwell noted that Ford was giving a fine target for the enemy sharpshooters. Cardwell called out to the lieutenant to dismount. Nevertheless Ford continued on horseback and simply replied that it was "as hot as Williamsburg," and then cried out, "Mack, you had better retire these guns." With the noise of the battery so deafening the captain could not hear his call. Ford then gave the order to limber to the rear, which was rapidly initiated. Cardwell described the action that followed:

> I was No. 4 at the gun next to the last to the right of the battery.... I put the pintle hook in my gun as she rolled off, and the horse holder brought my horse up to me.... just as I took seat in the saddle a Frenchman named Antonio called: "Garcon, Lieutenant Ford is killed." I looked and he was catching Ford's body as it fell from his horse.

Cardwell attempted to ride to catch the lieutenant to take him off the field, but just as he made his turn, a bullet found his mount and killed it. With the air full of missiles and the sky aglow with fire, Cardwell, dazed, recovered and jumped up behind Sergeant Temp Brown's horse to get off the field.

During the mad dash to the rear, Brown and Cardwell came upon a gun "whose trail had run up on a stump, and, the pintle hook not being in, the gun slipped off the pintle and was left behind." The two pushed the horse harder to catch up with the gun's limber and had it returned to recover the gun.

A short distance to the rear, Lieutenant Ford, with a bullet wound to the forehead, lived for nearly two hours before expiring. Dr. Kinney of Staunton, surgeon of the battery, tended to Ford until he breathed his last. Arriving in camp at 10 p.m., Lieutenant Brown rode to see General W. H. F. Lee to announce Ford's death. Lee responded by saying he was "very sad that we are losing so many of our gallant young officers. He complimented him very highly not only for his bravery, but for his coolness and judgement." There was little time for a formal burial. Lee said, "You will have to bury him where you are. We are to move at four o'clock and do not know our destination." Shortly afterward, Brown returned to the battery's camp where the men wrapped the young lieutenant's body in a blanket and laid him to rest in a fence corner Hanover County just after midnight.

Two days later, finding that the grave was back in friendly lines, Lieutenant Brown asked Reverend Joshua Peterkin to arrange for burial within Richmond. Peterkin had the lieutenant's body laid to rest in his personal lot in Hollywood Cemetery. Gathering a few men, Lieutenant Brown took an ambulance out to the site of the earlier internment, recovered the body, and placed it in a rough box for a casket because "there was such a demand [for caskets] at the time in Richmond." Upon their arrival in Richmond, Reverend Peterkin read a brief service and the men gave their gallant lieutenant a proper funeral. Lindsay, a faithful black servant, was then directed to return to the Ford home in Fairfax with the lieutenant's personal effects and a letter to his mother and father explaining their son's demise.

On June 1 McGregor's Battery was at Ashland, in position, but unlike the cavalry, was not engaged. The battery moved on June 3 and 4 to Meadows Bridge. F. T. Davis was wounded on June 4. The battery may have come into action again on June 7, as William Grey received a wound through the left hand on that day.

For a few days following the move to Meadows Bridge, efforts were made to improve the ordnance of the 2nd Stuart Horse Artillery. On June 10, General William N. Pendleton wrote to the Chief of Ordnance for the Army of Northern Virginia requesting replacements for the guns recently exchanged between General Long and McGregor's horse battery. The horse artillerists parted with two bronze Napoleons that were "too much worn, served, and honeycombed to be properly serviceable in the field." In exchange, two 3-inch rifles were added to the 2nd Stuart Horse Artillery's armament.

Within days of the receipt of their new guns, on June 12 at White Oak Swamp, McGregor's Battery was attacked and held its position for nearly four hours before being relieved by General John C. Breckinridge's infantry. The following day the battery moved to a position near Malvern Hill, where it went into bivouac for several days.

In the days that followed, Grant began to slip his men across the James River toward Petersburg. Still convinced that the Union general had his sights on Richmond, Lee was reluctant to shift to defend Petersburg. In the meantime, Beauregard remained in defense of the city with only 2,200 men. As fighting raged around the Cockade City for the next few days, Dearing's Brigade and the Petersburg Artillery were assigned to W. H. F. "Rooney" Lee's division. Graham's Battery and the 2nd Stuart Horse Artillery would share many experiences until the close of the war.

Leaving Malvern Hill, McGregor's Battery crossed the James and took position on the right, south of Petersburg. On June 21 a large cavalry force under General James H. Wilson and August V. Kautz struck the Petersburg and Weldon Railroad at Reams' Station, tearing up several miles of track. Wilson pressed on to Black's and White's (now Blackstone) on the South Side Railroad before being attacked by portions of Rooney Lee's division. Wilson attempted to drive off his assailants several times without success.

Rufus Barringer of the 1st North Carolina Cavalry wrote of June 21:

> ...while guarding the Petersburg & Weldon RR at the Davis Farm, below Petersburg, my pickets notified me of the approach of a large Yankee force of infantry.... we were wholly w/out support, but the thick undergrowth and other surroundings favored a vigorous resistance in a dismounted fight. I selected a high point for my horse artillery under McGregor, and as far as possible screened it from the enemy's view.... also kept the 5th VirginiaCavalry mounted in reserve to support McGregor... then dismounted the 1st, 2nd, & 3rd Cavalry and formed two heavy skirmish lines well concealed in thick undergrowth in front of the railroad with instructions for the first.... not to fire until the Federals were less than 100 yards of them & after a single volley to slowly retire on 2nd line, at this juncture four guns were to open - the plan worked well.

On the morning of 22nd the enemy passed by the right and moved in the rear before it could be intercepted. All day the Federals continued to move along the line of the railroad toward Danville. On June 23 the division caught up with the enemy near Black's and White's. At about 1 p.m. the 2nd Stuart Horse Artillery was deployed in an open field to the left of the road near the bridge, where it soon opened a hot and heavy fire to the immediate front. At one point the Federals had the dismounted Confederate cavalry on the run toward the guns. As Rooney Lee's North Carolina cavalry came up, Lee yelled out: "Save the guns! Save the guns!" Quickly the Tarheels deployed on foot and immediately moved to cover the guns. One Federal cavalryman was said to have fallen at the very mouth of one of McGregor's guns, knocked out by one of the gunners with a rammer. In a short while the fighting was taken back a distance to a timber line where it continued throughout the day. McGregor's guns could not have the same effect throughout the day due to the amount of timber on the range, but did take up a post next to the railroad and continued shelling. At the close of the fighting, McGregor found the colonel of the 1st North Carolina Cavalry, seized his hand and exclaimed, "Henceforth those guns belong to the First North Carolina Cavalry, you saved them today and they are yours."

Lieutenant Shreve, while mounted and in charge of a section during the engagement, was wounded severely and was cared for at the mansion of Freeman Epes. As it would turn out, at the close of the war Shreve would marry Mr. Epes' niece.

Though no action that day is of record, on June 30 Carter Irvine was wounded in two fingers on the left hand. He never returned to duty. James C. Hamilton was also wounded in the foot and at the General Hospital in Charlottesville on July 4.

In a follow-up inspection of artillery, Captain Cooke visited McGregor's Battery on July 4. Having been absent from the first inspection, McGregor's gunners like the other batteries had little time to prepare for such a call. Cooke wrote:

> The condition of the battery does not materially differ from that of the rest of the battalion. The horses are broken down by severe marching, and many of them are badly galled on back and shoulders; but they are evidently improving from the rest, and I think will be fit in a few days for active service again, unless movements are uncommonly rapid and arduous.
>
> These horses of the battalion generally are excellent animals and only require rest. They are properly groomed and get full feed of oats and about five pounds of corn.

> Captain MacGregor is using all his available transportation to haul oats.
>
> The guns and ammunition are in good order; the harness and accouterments, especially the saddles and bridles of the cannoneers, in bad order. The saddles were originally of the bad Jenifer pattern and being now nearly worn out are ruinous to the horses. All articles of this description were inspected and condemned and a certificate left for Lieutenant Boteler, ordnance officer, who will be able to procure the excellent new saddle of the Allegheny pattern now made by the Ordnance Department, and such harness, &c., as are needed to supply deficiencies. Thomson's and Hart's batteries have moved with the quartermaster's train to a point on the Dinwiddie Court-House road about ten miles from Petersburg. MacGregor's battery will probably move to the same place to-day or to-morrow.

On July 5, news reached McGregor's men of the fall of Atlanta. Spirits remained high, however, as the men feasted green corn. The lull in activity would carry through until the end of the month. By July 24 McGregor's Battery was reported at Armstrong's Farm.

Apart from the capture of C. Antonio (7/8), the wounding of F. H. Little (7/24), and John C. Phillips being killed in action (8/6), very little is recorded of the activities of the 2nd Stuart Horse Artillery until the last week of August. By that time, General Winfield S. Hancock's II Corps was moving against Reams' Station, on the Petersburg and Weldon Railroad. Late on August 24, Generals Wade Hampton, A. P. Hill, and Butler held a council of war and planned on meeting Hancock's offensive the next day. By midnight, "boots and saddles" was blown, and the cavalry prepared to move at once.

On August 25 Hampton moved with the divisions under Butler and Barringer to cooperate with A. P. Hill in the attack upon Reams' Station. Directed to attack the Petersburg and Weldon Railroad by Hill, Hampton moved with the bulk of his force to the left of the enemy. The remainder covered Hill's advance on Reams' Station. McGregor's Battery moved with Hampton and participated in the fighting which followed. The battle continued for two hours until 6 p.m. One of McGregor's 3-inch rifles was particularly effective in the fight, drawing praise from General W. H. F. Lee and Hampton. Shreve, manning the piece, wrote also that he had been "complimented on a number of good shots... made with a 3 in. rifle gun...by the officer in Command, who was close to the gun."

By the close of the fight the Confederates had captured seven pieces of artillery and a large number of prisoners. Of the service of McGregor's guns, Hampton commented that "by a rapid and well directed fire [they] contributed greatly to the confusion of the enemy.... Their guns were admirably served during the while engagement, and I beg to express my entire satisfaction at the conduct of Capt. McGregor and his men." A South Carolinian recalled:

> That day General Butler had some difficulty in restraining McGregor's Battery, which, with his guns, was stationed on Butler's line. McGregor asked permission to gallop up right in front of the Yankee line of battle, unlimber, and commence firing. Of course this could not be permitted.

For his "valor and skill" in action, Ashton Chichester was promoted to 4th sergeant on August 29.

All was not warfare for MacGregor's Battery during the hot summer months outside Petersburg. Caldwell recollected a time "...late in the summer of '64, when the cavalry was trying to hold the Yankees back from the Boydton plank road and the Weldon Railroad":

> ...One hot afternoon our battery was ordered into camp near Jones' farm. When we had spread our blankets and unsaddled - that was all we had to do, as we had no tents- some of the boys went out to see what the neighborhood afforded in the way of something to eat. Very soon one of them returned with the information that Mr. Jones' water-melon crop had not been harvested, but lay in full view, and that the ground was fairly covered with luscious melons. The fellow who brought the news was asked why he did not bring in a sample, when he informed us that there was a guard over the patch. For a moment the situation was gloomy. Soon Joe Pearl, rarely ever downed by trifles, said: "Boys, I've got it; get ready to eat Jones' melons. Do you fellers join me! I will go on guard in the field myself and you boys lay along the fence in the bushes and I'll roll out the melons." We had great faith in Pearl as a manager, but did not see how we could get away with the guard already in charge. Well, we did as he directed. With a sabre hanging to his belt he mounted the fence and boldly advanced on the "true, true" guard, who, in time, challenged, "Halt!" Guard Pearl did not stop, but commanded, "Halt!" "Who are you?" I am on guard here," he answered. "Whose command do you belong to?" demanded Pearl in haughty tones. He replied, to Graham's Battery. Ah! That's all right, said

Pearl, I am from MacGregor's; you watch that end and I will watch this, and the goose said "all right." In a few minutes our "guard" began to roll out the largest and best melons under the fence. We held a council of war to decide whether we should give our officers any of them or not. Finally we decided to roll a melon or two down to where they had their bivouac, and it was noticed that not one of them even cared where they come from. One was a preacher, too.

By the end of the summer, food supplies were dwindling. Rations already had been diminished a great deal before the men went into the works around Petersburg. David Cardwell of McGregor's Battery wrote:

> ...My comrades know how we were put to it for something to eat. Sometimes we had bread, such as it was, sometimes meat, and sometimes neither. Men resorted to all sorts of devices to get a square meal. If perchance they met a farmer, they at once cultivated him as a long-lost brother and made all sorts of excuses to call, took the girls to ride, etc., and never left without eating some meal, either dinner or supper. Our orderly sergeant, a Frenchman of many accomplishments, is said to have called on the widow Hancock in Dinwiddie County and on taking his leave also took her gray cat, and his mess ate her in a stew smothered with garlic the next day. "They say so;" I don't know. A Frenchman has the reputation of eating anything.

Realizing that the Federal army had on hand several thousand head of cattle, Captain George D. Shadburne, chief of Confederate scouts, hatched a daring raid to feed the army and suggested the plan to Wade Hampton. Shadburne spotted a herd of nearly 3,000 between City Point and Coggins Point on the James River. General Lee decided that Hampton should make a wide circuit in order to get the cattle out to safety in the Confederate lines. The route would take the Confederates 40 miles more than Hampton's original proposed route.

Before dawn on September 14, the Confederate cavalrymen mounted with four-days' rations. At the head of the column rode Wade Hampton with General Thomas L. Rosser's Laurel Brigade, followed by Dearing's Brigade and Rooney Lee's division. One section each from McGregor's and Graham's batteries rode along as a part of the artillery detachment. The wheels of the guns were ordered muffled silenced by means of grain sacks wrapped around them. Cardwell remembered:

> On the morning of the 14th of September, 1864, long before daylight, we were aroused from our camp by the notes of the bugle sounding "boots and saddle," and the command to which I belonged, the Stuart Horse Artillery, was ordered to saddle up and move out behind the 13th Virginia Cavalry. We waited, seated on our horses, for a long time (all waiting seems long), and while we were waiting we speculated upon where we were going and what we were going for....

Traveling along the Boydton Plank Road, the column shifted southeast on Flat Foot Road until within sight of Wilkinson's Bridge on Rowanty Creek and then bivouacked for the night.

The following morning the trek resumed two hours before sunrise. Heading this time in a northeasterly route, the cavalry force crossed the dismantled Norfolk and Petersburg Railroad by noon. Arriving at the destroyed Cook's Bridge over the Blackwater, Hampton deployed the engineers to span the river. The South Carolinian had been well aware of the bridge being destroyed and had decided on the route for that particular reason, figuring the Federals would not patrol the area as much. As the engineers laid 100 feet of bridge, the cavalrymen and horse artillerists lounged in the grass, feeding their horses and uprooting sweet potatoes to feed their hunger. Due to the constraints of the mission, campfires were not allowed and the sweet potatoes were eaten raw.

By dusk the bridge had been completed, allowing the column to continue across Blackwater River. Shreve remembered that "About midnight, we resumed our march, the road shaded by pines and in good condition. The night was lovely with a full moon." Once on the north side, the command was divided. Rooney Lee would march to the north on Lawyer Road toward Sycamore Church. Rosser would advance on Walls Road, east of and parallel to Lee in a direct line with Coggins Point. Lee was told to anticipate Federal pickets. Dearing would move along Hines Road on Rosser's right toward Cook's Mill on the Stage Road. When the guns from Rosser's command were heard, the attack would commence.

At 5 a.m. the attack by Rosser opened, driving in the 1st District of Columbia Cavalry. In short order, the Confederate columns converged and threw off the enemy horsemen and the 120-man guard around the herd of cattle. Shreve recalled that the Federals "were armed with the new Henry repeating rifles...the first that we had seen, and of course they excited our interest and curiosity. We fired a few shots on the fleeing troopers." Some 2,486 head of beef were in Confederate hands. About 8 a.m. the Confederate column began its withdrawal with the cattle, prisoners, wagons, and three stands of captured colors. Shreve recalled that "the men indulged in a little fun, calling on the foe to 'Come up and draw your beef.' We all had a good time, and relished the prospect of fresh beef for a while."

The withdrawal commenced as the attack had entered Coggins Point, in three columns. Though harassed by gunboat fire from the James River, the men continued moving at a trot instead of a gallop. The choice of speed lost valuable time and allowed an enemy response. Hampton, realizing the situation, sent members of his staff to drive the herd toward Hawkinsville while the main column moved toward Jerusalem Plank Road. Near Ebenezer Church, the Federals were only one mile behind them and closing. Rosser threw up hasty breastworks of fence rails behind a swamp and clashed with the enemy briefly before resuming the march. A member of the Petersburg Artillery later wrote that the "Yankees overtook us at Belcher's Mill, where a short and decisive battle was fought to regain the cattle...."

Though darkness and the lethargic nature of the cattle slowed the column even more, the Confederates splashed across the Nottoway River at Freeman's Ford on September 17 and back into their own lines. Only two men were recorded as captured from the 2nd Stuart Horse Artillery on September 17 as losses in the raid. Peter Paralli was one of them. While at Point Lookout, Paralli had a change of heart and joined the U.S. service on October 14.

The raid received numerous accolades. Hampton's feat was compared to J. E. B. Stuart's master stroke when he rode around McClellan's army in 1862. An editor for a Richmond newspaper described the harvest:

> The Federal commissaries buy beeves of the largest size for use of their armies in Virginia....the beeves taken in Hampton's late expedition are judged, by a London grazier, to weigh 800 pounds net. 2,486 beeves at 800 pounds would make an aggregate of 1,988,800 pounds, or within a fraction of 2,000,000 pounds. This distributed in daily rations of a pound each, would feed 1,000 men for nearly 2,000 days, 10,000 men for 200 days, or 50,000 for 40 days and so forth.

From September 14 through September 30, General Pendleton reported, Hart's Washington (South Carolina), McGregor's 2nd Stuart Horse, and Graham's Petersburg horse batteries, previously connected with General P. G. T. Beauregard's command, had participated in several spirited affairs of the cavalry on the right flank, under command of General Hampton.

While the Petersburg battery saw heavy action at Peebles Farm on September 29, no like action involved the 2nd Stuart Horse Artillery. Shreve wrote on October 26 that "since the last date, we have had no engagements, worth mentioned. A few false alarms have disturbed us occasionally."

For the next two months little of consequence affected the battery apart from the normal false alarms. On October 26, General Hampton held a review of the cavalry near Dinwiddie Court House and the battalion fired seven rounds in his honor. The following day the command marched in the morning toward Petersburg.

On October 27, McGregor's and Graham's batteries effectively cooperated in a combined attack that drove back the enemy in "confusion with a heavy loss." The enemy had pressed in the Confederate right. The Confederate cavalry in the meantime got in the rear of the enemy and on their left, finally checking the Federal advance by 3 p.m. Joining with the main Confederate infantry line, Rooney Lee's men pressed the enemy until dark. In the affair that day the batteries of McGregor, Hart, and Graham were all engaged. While checking the enemy advance upon the Boydton Plank Road, Hart received a severe wound which would cost him a leg. Ellett's and Gregg's batteries also participated in the fight.

For almost a week following the action of October 27, McGregor's men returned to picket duty. On October 30, Shreve recalled "Religious services around a huge camp fire." On November 8, the 2nd Stuart Horse Artillery, accompanying cavalry, left camp and marched up the Weldon Railroad. Halting in the road for the night the men anticipated moving at anytime. However, by daybreak the enemy had retired and the battery returned to camp.

For the balance of November, life was relatively quiet for the men of McGregor's Battery. Shreve recalled on November 9 that they had expected to hear of Abraham Lincoln's re-election. Otherwise, on November 16, President Davis had appointed a day "for prayer for our country." By the close of the month, however, the battery would once again be on the move.

While camped near Dinwiddie Court House, at 3 a.m. on November 30 the battery was put in the saddle and made ready to move, where they remained all day before dismounting and returning to camp at noon. Two days later, while on drill, the battery was again prepared to move, and this time did get on the road toward Stoney Creek Depot. General David M. Gregg's Federal division of cavalry had struck the Confederate pickets at Rowanty Bridge. In an effort to aid the detachment holding the depot, the column marched 12 miles before it arrived too late to help. Shreve wrote that on their trip to Stoney, they "were without rations the whole time."

In early December, McGregor's Battery did little but drilling. On December 7, however, there was yet another raid that the Confederate cavalry turned out to resist.

In response to the enemy movement in force down the Jerusalem Plank Road, Rooney Lee moved on December 8 with his division and McGregor's Battery to Stoney Creek, arriving there shortly after daybreak. The Confederates crossed the Nottoway River at Gee's Ford and engaged the enemy near Field's House for about two hours. Finding that the enemy was moving a strong column on the road in the direction of the fighting, Rooney Lee's and Butler's divisions recrossed the river and moved twelve miles to the right on the enemy flank. Bivouacking for the night in a "cold piercing wind," the men roasted shelled corn in hot ashes for the morning's breakfast. The following day the march was resumed at 2 a.m. The weather was so cold the men were forced to dismount at intervals to walk and maintain circulation. By 2 p.m. the men reached Belfield, where the enemy was again encountered. The battery along with Hart's opened the fight and kept up the firing until dark. The march was then resumed in a freezing rain and without rations or shelter. The following morning the enemy had withdrawn.

The batteries were complimented in an official report: "Their guns were effectual in repelling the enemy at Hicksford and admonishing him speedily to retrace his steps." John Finks may have been wounded in the action.

While winter continued to blow in, the 2nd Stuart Horse Artillery moved into winter quarters at Belefield, in Greensville County. The men hoped the four 3-inch rifles of the battery would have the chance to remain cool for a while. It would be no comparison to the more bountiful times of the previous winter at Charlottesville.

Chapter 8
1865
1st Stuart Horse Artillery

Three days into the new year rumors flew about the camp concerning possible disbandment due to lack of forage and rations. On Sunday, January 15, the batteries of Breathed's battalion were disbanded for the winter. However, some muster records record that the 1st Stuart Horse may have continued active. T. H. E. Elaim and John Llewellen were recorded as captured on March 7 at Mount Jackson and March 2 at Waynesboro respectively. On March 11 there was a desertion when John Fay slipped away.

Though Breathed's battalion still existed on paper and was headquartered at Waynesboro, most of the men returned to their homes or went to friends' homes with the understanding that they might be called into service again before spring. Upon reorganization of the battalions on March 20, the 1st Stuart Horse Artillery, now under Captain Daniel Shanks, with the promotion of Johnston to major, was placed in Johnston's own battalion, along with Thomas E. Jackson's Kanawha Battery. This new two battery battalion was designated to serve with Lunsford Lomax's Division. Jackson's Battery was at that time encamped on Walker Creek in Giles County.

Whatever happened to the 1st Stuart Horse Artillery beyond March is left to speculation. While relatively few men from the other horse artillery batteries were back in the field fighting at Petersburg and Appomattox, Shanks' Battery leaves no trace. The other battery in Johnston's battalion, the Kanawha Battery, did receive word to report by early April. However, while it was marching for Lynchburg, news was received of General Lee's surrender.

Major Breathed did make it to the army prior to Appomattox, and was nearly killed at High Bridge when James Thomson, of the Ashby Artillery, and Brigadier General Dearing were killed. What few men accompanied him to that point is uncertain. James F. Brown was the only man recorded as surrendered at Appomattox on April 9. Breathed received his parole at Winchester on April 24. At least one man, Dennis Conroy, made his way south and was paroled at Greensboro on April 26.

After a glorious history in the field and an unblemished war record, the men could proudly return to their homes amidst a difficult time in trying to return to normal lives.

1865
2nd Stuart Horse Artillery

Inactivity in winter quarters came to an end in March. Moving from their winter quarters, by March 15, the 2nd Stuart Horse Artillery was in camp at Dinwiddie Court House. As the pressure on Petersburg increased, so did the activity of McGregor's Battery. The last few days were best recounted by the stories of Private "Davy" Cardwell and Sergeant George W. Shreve. Cardwell wrote:

> On March 31 the cavalry found itself way down the line towards Dinwiddie Court House and they put their artillery in position on the south bank of a stream called Chamberlain's Run (I was No. 4 at one of these guns) and in a few minutes the First North Carolina cavalry in the lead (a splendid regiment), under Col. W. H. H. Cowles came up and Barringer's brigade. At once a heavy fire was opened on them from the other side of the run. We opened with our guns, using short fuses, and as the brigade of Gen. Barringer charged across the run, which was about three or four feet deep, our gun was in position on a little knoll on the right of the ford by which the cavalry charged and I had a splendid opportunity to see the whole fight. In all my experience, and I had been in over 60 fights, great and small, I never saw a more splendid charge; they simply swept everything out of the way.... until they reached Dinwiddie Court House, where their infantry was in large force.... One of our men, a splendid soldier, Ashton Chichester, was shot through the thigh....

Following the fight at Chamberlain's Run, Cardwell commented that "we thought we could whip anything alive." Of the conditions of the men and what followed, he wrote:

> We had taken care of our wounded, buried our dead, and were ready to go to sleep. We went to sleep and slept as only soldiers can. The matter of bed clothes did not concern us, we "traveled light" and when reveille sounded at dawn April 1, it was soon followed by "boots and saddles." We were up and in the saddle after a hasty breakfast of...corn pone cooked three days before and raw Nassau pork, sometimes called "mule" by the boys who worshipped it and got so little of it. I was hungry--hungry is not strong enough--I was so hungry that I thanked God that I had a backbone for my stomach to lean up against....

Map, Five Forks, April 1, 1865

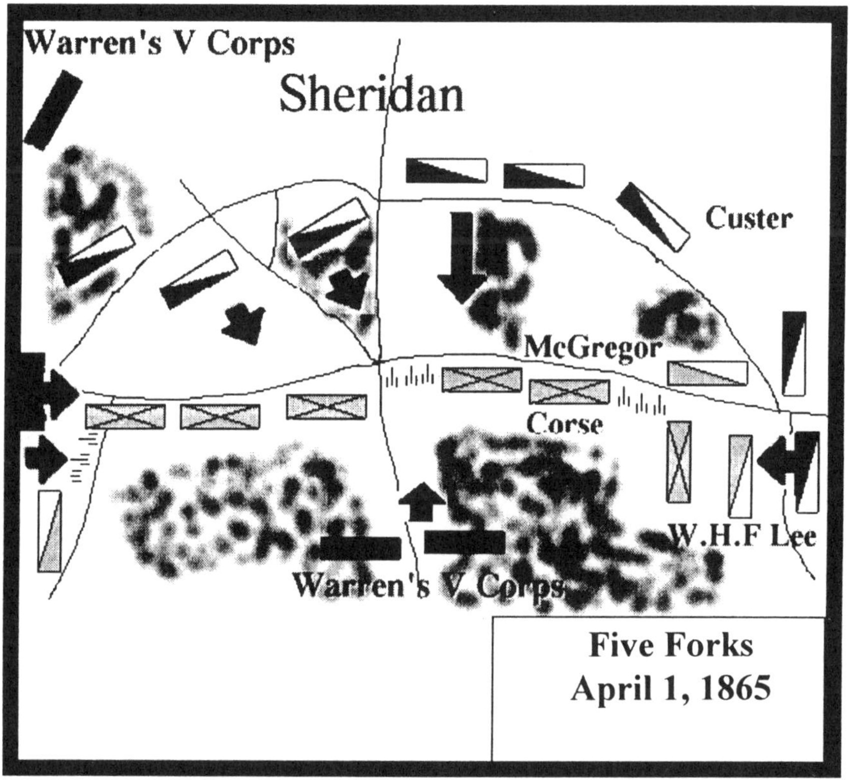

We found later that we were on the extreme left of a line that was forming and we found this out by being ordered to the extreme right at a gallop. As we galloped down the line we recognized among the infantry Pickett's men, Bushrod Johnson's men, and the foot artillery that belonged to those commands. We passed just in the rear of them all. They were behind very light earth works--just enough to insure them a wound above the knee if they got one.

We recognized many men in the infantry and artillery that we knew, many who never left that field alive. One man in particular I noticed sitting on his horse just in the forks of the road (Five Forks). He had on a brand new uniform of gray, with a red collar and on the collar three gold stars. He was a young man. (Oh! So young he seems to us old soldiers now!) To my boyish eye (I was 19) he was simply grand. Think of it, he was a full colonel of artillery! Could anything be more exalted than that! He had on gold spectacles and looked to me the god of war. It was Col. "Willie" Pegram of old Virginia....

We galloped down the entire line and on the right we found our division (W. H. F. Lee's) already dismounted and in position.... behind breastworks.... Space was made for us at once and our four guns went into position. "On the left, 14 yards, close intervals, forward into battery!" was about the command our captain (Wilmer Brown) gave, and we unlimbered for one of the history-making battles of the world.... If anybody said anything it was about like this: "Well, boys, looks like we are going to have a good size scrimmage."

We dismounted; gave our horses to the horse-holder; proceeded to rest ourselves and to make remarks. One said: "No fight today, boys; only an April fool." Our commissary rode down the line on a good fat horse and such cries arose as, "No fight today, Company Q is at the front." Another fellow would sing out, "Look at that horse, how fat he is; that's where the feed goes" and so on. The careless jolly humor of the men almost in the presence of death is hard to understand in these piping times of peace.

Well we lounged about these works till say about 3 o'clock. At one time one of the boys went back to the limber chest and got out of it a plug of tobacco so musty that no one could use it--such as our government issued to the troops. He suggested that we ram it into the gun and said: "If anything will kill them, this will."

About 2:30 or possibly later some one said, "I see some movement in the peach orchard back of Mrs. Gillam's house," and as I could not see it (being a short man) I went back and got my horse and as I rode up, Major McGregor, who was formerly our captain, handed me his field glasses and I beheld a sight! Such a sight is rarely vouchsafed to any one more than once in this world. I saw a squadron after squadron of splendidly mounted and equipped cavalry wheel into line in the orchard back of the Gillam house. They were at a "carry sabre" and were forming by the right flank. The sun shone brightly and the glint from their burnished arms was blinding. On they came, forming on the right-right in front of our position. Well, I watched them until I was sure there was more cavalry than I ever saw; even on the plains of Brandy Station.

I saw an ambulance drawn by four horses drive up to the Gillam house, and an officer in charge (I don't know whether Yankee or Confederate) assisted the ladies of the Gillam family in and drive off. I thought by this time I'd better get to my post, so I gave up my horse and went to my place - No. 4 at the gun....

After a while, out in front we could see a small white puff of smoke, then another, and so on, till it sounded like a pack of firecrackers... And raising our eyes we saw them coming, at first at a walk, then a "trot march" and next "gallop-charge" and then bugles blew the charge.... We threw short fuse shells into them as fast as we could pull the lanyard and then canister. On they came! We though they must ride over us, but when the canister got in its work they wavered swayed and turned back....

We looked out over that great field as the smoke cleared away, and saw the still bodies of the dead; the slightly wounded crawling off, afraid to raise their heads; the poor horses some still in death, some struggling to get up. This was repeated four times (as I remember) and when they came the fourth time we heard the spattering of bullets in our rear to the left. In a few minutes we saw the infantry moving towards us; first just one or two, then three or four and so on till there was a big movement through our command and then a stampede.

Our division kept its line and never wavered, though the infantry crowded them. They met the charges and after all-moved out under orders. It was an awful mixture. Gen. W. H. F. Lee took a flag and rode among them begging them to rally, but no sir, they would not--they went on their way.

We limbered up under order to do so, and moved out toward the right - down the road. As I attempted to follow with two other comrades we were stopped by... Col. L. Tiernan Brien.... He told us to pull down a fence in front of the Second North Carolina cavalry "squadron front," and we did it. By so doing we got (after we had mounted) terribly mixed up in a charge of that regiment. We were pressed back to the road and I was looking for a way out and I was looking hard, too.

I looked west and there was a line of fire clear across the road. I knew I could not go east. The enemy were there, and south. So northwest I went. I jumped my horse over a fence. Poor old Dick Chamberlaine (of Norfolk), who was No. 2 at my gun had all of his eyebrows, mustache and hair on the right side of his head burnt off. Half of his jacket was also burnt off by his being too slow to clear the front of the gun when the order to fire was given. Poor old Dick, my messmate, rode close behind me, and jumped the fence into the swamp also. Dick was a sight. You could not tell which army he belonged to. Your humble servant was also a sight. My face, I found afterwards was black with powder, and when I woke up from my excitement I had my attentions called to it.

The Second North Carolina cavalry had checked Custer's mad charge which enabled the artillery and most of the cavalry to get away.

Henry L. Moore of our battery, Owen Brian (Pat), and myself were close together. In the mix up a color bearer was cut away from his horse, and as the flag fell, it was seized by Moore. A great big man made a swipe at Moore with his saber, missed his head and struck the flagstaff. Moore struck him over the head with his pistol. Moore brought the flag out.... This about closed the battle of Five Forks. Night had come on before we got through, and the line of fight could easily be traced by the line of fire. Soon after we got in the swamp we encountered the numerous wounded.... As we rode along they begged piteously to be taken up. We talked about it, but could not see that we could do them any good. We know that the ambulance corps of one army or the other would look after them in the morning.

I found I had, when I jumped the fence, a saber hanging to my wrist by a strap and a pistol in my hand. With Dick running by my side, I looked much like a man with a prisoner. Soon a man said to me, "Great God man, how did you get a prisoner?" I put up my saber and pistol and looked more peaceful. Dick and I floundered in that swamp and soon were out of sight and sound of the wounded and all sound of strife.

We keep on in a northwesterly direction and just as the roosters would begin to crow (if I dared to do so) we emerged into an open field. We halted and in a few minutes we discovered the figure of a man standing upright holding a gun. He looked like "one in authority." In fact, he looked "on guard." So we contemplated him with interest, Dick and I. Dick says "Davy, you ride forward and challenge him." I said, "Dick, suppose you amble up and ask his name and command." We did not know where we were, nor who he was. After some more debate we both rode forward and the figure promptly presented his gun and said "H-A-L-T" in a cross tone of voice and we halted as fast as we could. We found he was on guard and belonged to our division. We asked him where our battery was and he said: "Way over beyond that big white house." And for over beyond the big white house we started.

Before we had gone far, however, we came upon a group of mounted men and in their midst we soon made out Maj. Gen. William Henry Fitzhugh Lee. It was now broad daylight and we could see things plainly. But I must say that in the darkest part of the forest, at the darkest period of night, we could have made out Gen. Lee. He was six feet three or four and weighed not an ounce less than 300 pounds. He was mounted on his brown horse Frantic, a horse he always had his saddle changed to, when practicable, before going into a fight....

> We halted here. Meanwhile Moore and O'Brien had joined us, Moore still carrying the battle flag. We found Gen. Lee trying to gather together and get some shape his scattered division....
>
> Now our turn had come. The general turned in his saddle and in a sweet, soft, courteous voice - (like his illustrious father) - he said, seeing the red trimmings on our jackets: "Well, young gentlemen of the artillery, what is your battery?" We answered, as we always did, with great pride, "McGregor's!" "Well, permit me to inquire why were you not with your battery?" We thought we had a splendid excuse. In fact, we felt that we had done even more than our duty.... Henry Moore rode up (with some style), and expecting to be complimented, showed Gen. Lee the recovered flag. We told him of Col. O'Brien and the line of fire to the west across the road. As our battery retired that way we thought it was captured. He answered, with more harshness than I thought him capable of, "Go to your battery; it is not captured, it is over there in that field. You should never allow yourselves to be separated from your battery." We rode off in no good humor and said about this: "He is a doggoned ungrateful old fellow and I don't like him anyway. He ain't much of a general...." We rejoined our battery in the big field and were received with great rejoicing.

From Five Forks, McGregor's Battery continued to follow Lee's line of retreat from Petersburg. At least six men were listed as captured on the route at Aberdeen Church on April 3: George Hicks, John F. Lewis, Sergeant Owen O'Brien, George A. Pryme, Joseph Spicer and John Strong. Pryme, a former 12th Mississippi infantryman, had a long way to return home and must have dreaded what he would find there in Vicksburg, Mississippi.

Of the long march, beginning on April 2, Shreve reported carefully every detail of the journey:

> ....while in the rear of one of our retreating columns, the enemy appeared close to us, marching parallel and south of our route. With no orders to engage, Maj. McGregor could not resist the temptation and so ordered us to pull down the fence, skirting the road, over which we took the guns and opened fire, creating quite a confusion in the ranks of the enemy, who notwithstanding our exposed position failed to return fire. McGregor was soon ordered to cease firing, and

we limbered to the rear, and proceeded along the same road, in full view of the Federals, marching on a parallel line. We could see the other plainly, as we marched peacefully along.

On that, or the next evening, near sunset, another temptation presented itself, when we were brought face to face with the enemy. One of our Company officers ordered us "In Battery," and we commenced firing, which seemed to be our proper thing to do; when a staff officer rode up furiously, and demanded to know by whose authority we were firing, and ordered us to cease forthwith, which we reluctantly complied with. We had always been accustomed to pitch in and help out in all critical situations, so it seemed strange that we were not allowed to do our best at this time.

There was no demoralization whatever in our command, all through the retreat, and up to the day of surrender we were in readiness to fight on the instant, day and night. One day, overtaking and passing through the columns of one of our infantry divisions, we being all in good order, and having been engaged an hour or so, previously on the occasion when we were ordered to cease firing, we were accosted by the tired infantrymen as we passed them, in somewhat the following words: "You fellows can do all the fighting you want; We've had enough. We've done our share; we've quit; no more fighting for us." Hearing these words from such brave veterans, who had proved their bravery on so many fields, they fell upon our ears as the death knell to our cause, which it proved to be. We, (our Battery) did not know that we were whipped, and we were ready to fight on just the same as in the past; but these men had borne the heat of battle in many desperate struggles, knew more than we, and were justified in their decision.

The last position taken in our Battery, was on the 8th day of April, at about 4 p.m., we were following a large wagon train, and coming to a place where a country road or trail intersected the main road, from the direction of the enemy, (south) Capt. Brown, was ordered out on this road to protect the train. The order was given by some superior officer. Capt. Brown posted one section about a quarter of a mile down this road. The section was under command of Lieut. Shreve; the writer had charge of one detach-ment, and I failed to remember the name of the sergeant in charge of the other. The small road led through the woods, and we were unlimbered and ready for action. All was still in front and we could see nothing and the silence of waiting gave us time to think. We discovered we had no videttes out, and no support on either flank, nor in rear, and so the situation was discussed pro and con. Capt. appealed to the Lieutenant for his opinion, not caring to express his honest convictions, and so evade a direct answer. Then he appealed to the writer. I replied that in my judgement anything that we could do would be futile, that we had always done our duty, and while there was hope we were willing to fight, but that in our present position, if the enemy advanced his sharpshooters would flank us, and we were helpless and furthermore, I considered that every life lost from now on, would be a useless sacrifice. Others had joined in the conversation.

Capt. Brown, said "I agree with Sergt. Shreve. 'Limber to the rear'." We fell back to the main road, which was full of wagons, and Gen. Lomax with his cavalry division coming up, we were ordered to follow the same. We marched by a country road leading north and west, towards the James River for several miles, and bivouacked for the night. Early on the morning of April 9th, we resumed the march in the direction of Lynchburg, passing on country roads, leading through farms, and not striking the main road until near Lynchburg, where we arrived about 5 p.m., and parked our guns in the outskirts of town. At once we were ordered to assemble in ranks, and an order was read to us, announcing the surrender, by General Lee on the same morning, including our command. We were dismissed from ranks for the last time.

Within a half hour, Major McGregor, mounted with other officers, beside him, announced that he was going to start immediately southwest to join General Johns[t]on's army and invited all who felt so disposed to join him forthwith. Quite a number fell in, and the party started off about dark. Many pathetic incidents occurred. Men were crying being overcome with emotion, produced by the thought of a crushing defeat, and a doubtful future, whether they should return home, or essay to join the expedition to Johns[t]on. A young man, whom I esteemed highly, was greatly agitated, and asked me what he should do. I told him that brother and I were going with the expedition, but that knowing him as I did I advised him to return home, that we

> were entering upon what would be most probably a fruitless mission.
>
> McGregor called out, "Fall in at once, all who are going with me." My brother replied, "Go ahead. We will overtake you," being engaged just then in writing a missive to his sweetheart, standing, bridle in hand. I was also ready waiting to accompany him. Very soon he and I started off, through the city, in pursuit of McGregor, and party. Suffice it to say we never overtook them, but came up with some who had dropped out of the ranks. Being tired and hungry, we rode up to a farm house, where they received from us the first news of the surrender, and gave us abundant refreshment. We continued our march, and halted and rested under some trees for the remainder of the night. The next day continuing our southwest course, we reached the vicinity of Salem, where receiving news of the surrender of Gen. Echol's command we halted and relinquished our purpose of proceeding further. After a rest of a week, we started for our home in Fairfax County. We went by way of Staunton, and Winchester, near which latter town we met the first picket of the enemy. They treated us with great kindness, and escorted us to the office of the Provost Marshall, not as prisoners, but as companions in arms. By direction of this office, we went to Millwood, the next day, and received our paroles.

Only three men, George and James Baughn, and Jesse Irvine, had surrendered at Appomattox. Lieutenant Edward Croyean waited until May 12 to surrender himself at Danville. Like their comrades of the 1st Stuart Horse Artillery, a long and noble history, unblemished by shame, and laden with merit and praise in the annals of artillery history had drawn to a close.

David Cardwell later wrote some years after the war as fellow comrades left the ranks of the living to join the men long since passed in the war:

> In the horse artillery some splendid men served and died: Pelham, Breathed, Henry, McGregor, Ford, Chew, Wilmer Brown, Shumaker [sic], Bamberg, Halsey, Johnston, Thompson, Burwell, Hoxton, Shreve, Croyeau—all gone, or nearly so, since the bugles sang truce. No more "Boots and saddle!" No more starvation and glory! God rest their souls! These men, as was said by General Stuart of Pelham, had cast their eyes over the battle fields of Virginia, Maryland, and Pennsylvania. Who of the cavalry corps does not re-

member them at Brandy Station, June 9, 1863, when they charged with the cavalry, fought on the skirmish line, and broke the record in the handling of field pieces?

Gen. Theodore Garnett once said there might be a question as to who fired the first shot of the war and who fired the last shot, but there could be no question as to who fired the most rapid shots. It was the horse artillery of Stuart's Corps, Army of Northern Virginia!

APPENDIX

Note: Due to the inconsistent and incomplete nature of the actual existing muster rolls, statistics would not be accurate. Therefore, they are not included in this volume.

Commissioned and Noncommissioned Officers:

1st Stuart Horse Artillery

Pelham, John - Captain
Breathed, James W. - Captain
Johnston, Philip P. - Captain
Shanks, Daniel - Captain
Cosgrove, T. William - Lieutenant
Febry, Moses - Lieutenant
Shaw, Emmett - 1st Lieutenant
Hill, Edgar - 2nd Lieutenant
Hoxton, William - 2nd Lieutenant
Shepherd, James T. - 2nd Lieutenant
Wigfall, F. H. - 2nd Lieutenant
Murray, Sterling - 1st Sergeant
Taliaferro, W. G. - Quartermaster Sergeant
Goodman, B. W. - Sergeant
Hobson, Dean - Sergeant
Humfries, J. R. - Sergeant
King, Edward A. - Sergeant
Russell, Thomas A. - Sergeant
Seymour, Charles L. - Sergeant
Smith, H. R. - Sergeant
Swancoat, William - Sergeant
Thomas, J. Henry - Sergeant
Watkins, W. C. - Sergeant
Williams, Z. F. - Sergeant
Coode, Demetrius - Corporal
Fay, John - Corporal
Gibson, Fayette - Corporal
Higgins, James L. - Corporal
Hopkins, H. H. - Corporal
Russell, Elijah Tarleton - Corporal
Terry, George R. - Corporal
Tonge, Richard - Corporal
Warro, Joseph - Corporal
Mackall, Robert - Guidon
Willis, Thomas E. - Bugler
Connor, Jonathan - captain of forge
Conroy, Dennis - Pelham's Orderly

2nd Stuart Horse Artillery

Henry, Mathias W. - Captain.
McGregor, William M. - Captain
Brown, George W. - Captain
Elston, William C. - 1st Lieutenant
Ford, Charles E. - 1st Lieutenant
Shreve, Richard S. - 1st Lieutenant
Burwell, Robert P. P. - 2nd Lieutenant
Croyean, Edward - 2nd Lieutenant
Chichester, Daniel M. - Battalion Sergeant Major
Dabney, Walter S. - Battalion Ordnance Sergeant
Guillemotte, Charles J. P. - Orderly Sergeant
Butterly, Thomas - Quartermaster Sergeant
Brown, T. J. - Sergeant
Chichester, Ashton M. - Sergeant
Cook, John F. - Sergeant
Holmes, Joseph - Sergeant
Lumpkin, John B. - Sergeant
O'Brien, Owen - Sergeant
Shirley, C. P. - Sergeant
Shreve, George W. - Sergeant
Wharton, William T. - Sergeant
Brown, James F. - Corporal
Brown, James T. - Corporal
Pryme, George A. - Corporal
Romain, Alexander - Corporal
Scarborough, William - Corporal
Byron, Reuben - Bugler
Smith, John C. - Bugler

The "Presumed" Guidon of the 1st Stuart Horse Artillery: Guidon or Personal Flag?

The guidon or pennant shown in this volume is an unusual piece and deserves special note. Designed in a basic swallowtail pattern, the guidon, measuring at only 10" X 16½", is essentially patterned after the First National flag of 1861 with a few exceptions.

The first noticeable feature (which cannot be clearly seen in a black and white photo) is that the horizontal bars are red, white and blue and not red, white and red like the First National. The possible explanation behind the difference may be found in the fact that some of the men in the Stuart Horse Artillery were of French or Creole descent. However, most, if not all, of these men later become part of the 2nd Stuart Horse Artillery upon the division of the original Stuart Horse Battery. Nevertheless, the guidon uses colors commensurate with the French National Flag. If rotated 90 degrees to the left, the colors of the guidon/pennant would resemble that of the French National. Therefore, it appears that the designer manufactured the flag with the intention to merge some major features of both the French National and the Confederate First National.

Furthermore, the eleven stars or mullets are four pointed, placed in the blue field which would permit the addition of two stars, perhaps with the hope that Missouri and Kentucky could be represented upon officially entering the Confederacy.

Other color patterns that are significant include the "1st" and "Horse Artillery" in yellow, representing cavalry; and the "Stuart" in red, representing artillery.

While the 27 pieces of fabric and materials show construction in the 19th century, the guidon remains a mystery as to its actual use. The size is not of prescribed artillery pennant size and may therefore, have actually been one battery member's personal flag. It was not uncommon for a soldier to carry a small flag of his own through service with a unit for sentimental value in later years. Showing no damage from powder, fragment or bullets, the guidon shows only light soiling and leaves the impression that it was not actively used.

Other possible uses of the pennant may have been unit marker, camp, drill or parade guidon. Having originally surfaced in a New Jersey estate auction, the guidon leaves the possibility that it may have been a captured relic of the war. The only possibility of capture from the unit as a whole existed in the 1864 Shenandoah Valley Campaign, where the 1st Stuart Horse saw extreme action that resulted in major loss of men, guns and equipment. Individual captures were more wide spread since the formation of the battery in 1862. Ultimately, with no personal identification marks upon the guidon, the mystery will go unsolved until further evidence surfaces.

Lieutenant Colonel John Pelham
Captain of the Stuart Horse Artillery and the Battalion's first commander.

Major James Breathed

Captain Mathias Winston Henry

Courtesy of the U.S. Military Academy

Captain Robert Franklin Beckham

Courtesy of the U.S. Military Academy

David Cardwell

Major Robert F. Beckham

James G. Dearing
The "would-be" 3rd Commander of the Stuart Horse Artillery Battalion

Lt. Col. Roger Preston Chew
Courtesy of VMI

Captain William R. Lyman
Adjutant

Joseph Henry Phlegar

Campbell Murdock

Courtesy of the Museum of the Confederacy

John J. Busic

Sergeant Major Otho Scott Lee

Courtesy of the Library of Congress

"The Bower" Frequented in the fall of 1862 by Pelham and the Officers of the Stuart Horse Artillery.

Sturart Horse Artillery Marker
Gettysburg, Pennsylvania

Monument to Major John Pelham
Kelly's Ford, Virginia

Courtesy of Eric D. Feid

Stuart Horse Artillery Battalion Flag

Courtesy of The Jefferson County Museum

A "Presumed" Pennant of the 1st Stuart Horse Artillery

Courtesy of Bill Strong, Jr., Baltimore, Maryland

ARMY OF NORTHERN VIRGINIA
STUART'S HORSE ARTILLERY
BECKHAM'S BATTALION BREATHED'S BATTERY
FOUR 3 INCH RIFLES
JULY 3 THE BATTERY ARRIVED HERE ABOUT 2 P. M. AND TOOK AN ACTIVE PART IN THE FIGHT UNTIL ITS AMPLE SUPPLY OF AMMUNITION RECEIVED IN THE FORENOON ON THE WAY HERE WAS EXHAUSTED. IT WAS WITHDRAWN FROM THE FIELD ABOUT DARK
LOSSES KILLED 6 WOUNDED 8 TOTAL 14

ARMY OF NORTHERN VIRGINIA
STUART'S HORSE ARTILLERY
BECKHAM'S BATTALION McGREGOR'S BATTERY
TWO NAPOLEONS AND TWO 3 INCH RIFLES

Frank Vizetelly's sketch of what may be Pelham's Battery
Near Kearneysville, October 16, 1862

Courtesy of Houghton Library, Harvard University

A Note Regarding the Rosters

The rosters that follow were extracted from information in the Compiled Service Records, along with various postwar rosters updated by veterans. When available, birth, death, and sometimes marriage dates and activities have been included. As far as the actual military records are concerned, all information has been copied with the exception of pay and in some instances clothing issue information. Sometimes it will be noted within the rosters that a veteran may have no more record available than "enl. 9/1/63. Pres. - 4/64." This usually suggests that either there was no more information available on the veteran from the service record or simply that nothing of particular note, personally, occurred. This is not to say that the veteran was not actively participating in movements, engagements, etc. Sickness, hospitalizations, wounds, etc. have all been included in the rosters.

When dates are given in parenthesis, it will indicate that the dates are approximate based upon when an event was recorded in the service record. For example, when a person was described as AWOL "(10/31/64)" it simply shows that the date of the AWOL was noted in the muster rolls on 10/31/64, not that he went AWOL on that date.

It is interesting to note that the rosters that follow may appear quite incomplete. However, it appears that the officers of these two batteries did not have a knack for keeping muster records current. Some enlistments are only annotated with the month and day, leaving the year a mystery, but suspect to 1863 or 1864 is the most completely documented year for the two batteries of Stuart Horse Artillery. Whether there was better record keeping that year, or the fact that these records were not among those lost in the evacuation from Richmond, we cannot be certain. Nevertheless, while reading through these rosters, the reader should remember that these two batteries were probably among the most active in the entire Army of Northern Virginia, and owing to that activity, probably did not have the time to maintain records as well as the foot artillery.

Abbreviations

AWOL - Absent without leave
B. Or Bn. - Born
Bttn. - Battalion
Capt./capt. - Captain or captured (distinguishable by the case of the letter "C")
CCH - Culpeper Court House

C.H. - Court House
Cav. - Cavalry
Chimbo. - Chimborazo Hospital, Richmond, Va.
CM or GCM - Court Martial/General Court Martial
Co. - Company or County
Cpl. - Corporal
D. - Died
Enl. - Enlisted
Ft. - Fort
GH - General Hospital
GO - General Order
GSW - Gunshot wound
Inf. - Infantry
KIA - Killed in Action
Lt. - Lieutenant or Light
Maj. - Major
M. - Married
MWIA - Mortally Wounded in Action
NFR - No Further Record
OCH - Orange Court House
Ord. - Ordnance
Pt. - Point
Pvt. - Private
QM - Quartermaster
Res. - Residence
SHA - Stuart Horse Artillery
Sgt. - Sergeant
SO - Special Order
Surg. Cert. - Surgeon's Certificate
UVA - University of Virginia
VMI - Virginia Military Institute
WIA - Wounded in Action

PRISONER OF WAR CAMPS

Camp Chase - Camp Chase, Ohio
Camp Morton - Camp Morton, Indiana
Elmira - Elmira, New York
Fort Delaware - Fort Delaware, Delaware
Fort McHenry - Fort McHenry, Maryland
Johnson's Island - Johnson's Island, Ohio
Newport News - Newport News, Virginia
Old Capitol - Old Capitol Prison, Washington, D.C.
Point Lookout - Point Lookout, Maryland
Rock Island - Rock Island, Illinois
Wheeling - Wheeling, West Virginia

The 1st and 2nd Stuart Horse Artillery:

The Rosters

ADAMS, JESSE A.: 1st SHA. Pvt. Enl. 2/1/62 in Alabama. Another source indicates enlisted 6/10/63. Pres. (11/2/63) Absent sick with epilepsy at CSA GH Charlottesville 3/14/64-3/31/64. On register of payments from 5/1-8/31/64. Pres. 7/24/64, 8/9/64, 11/31/64. Paid $67.13 on 9/5/64. Absent sick 1/20/65. Another note in file indicates enl at Richmond 7/12/? by Lt. J. T. Shepherd. Paid $50.00 bounty. Postwar note indicates wounded at Amesville.

ADDISON, WALTER DULANEY: 1st SHA. Pvt. Enl. 8/1/61 in Richmond? Another note states enlistment on 7/?/63 at Richmond by Lt. Johnston. Present 11/2/63. Present 4/1/64. Among those who wished to transfer to Md. Line per S.O. #100/2 4/29/64. Captured 6/11/64 at Louisa CH. Another card states captured in Wilderness 5/8/64. Sent to Belle Plain 5/17/64 and then Pt. Lookout. On rolls at Elmira, N.Y. Paroled and sent to James River for exchange. On roll of exchanged POW's at Camp Lee 2/2/65. Paroled by 20th N.Y. Cav., Richmond 5/13/65. See POW cards in case of Frank Dyer, 6th Va. Cav and Theodore W. Edwards 15th Va. Cav. Res. of Price Georges Co., MD.

ADKINS, P.: 1st SHA. Pvt. Enl. at Halifax by Capt. McPhail. Enl. with SHA 1/28/63 from Camp Lee. At Chimbo. #4 with hernia 5/27/64. Furloughed 35 days. 8/30/64 to Halifax Co. Present (9/64). NFR.

AIKEN, JOHN E.: 1st SHA. Pvt. From Baltimore. Enl. 9/8/62 in Richmond by Lt. Shepherd. Present (11/2/63). Absent on furlough (1-2/64). AWOL (4/1/64). Postwar note states killed at Dover, near Aldie.

ALDRIDGE, STAUNTON R.: 1st SHA. Listed in Amos D. Woods' *Floyd County: A History of Its People and Places* as a member of this unit. Age 40, merchant, 1860 Floyd Co. Census. May have also served in the 42nd VA Inf., and the 37th Battn. VA Cav.

ALLEY, ANDERSON: 1st SHA. Listed in Amos D. Woods' *Floyd County: A History of Its People and Places* as a member of this unit. Age 40, farmer, native of KY, 1860 Floyd Co. Census. Also served in the 4th VA Reserves.

ALTIZER, ENNIS: 1st SHA. Listed in Amos D. Woods' *Floyd County: A History of Its People and Places* as a member of this unit.

ALTIZER, JOHN W.: 1st SHA. Listed in Amos D. Woods' *Floyd County: A History of Its People and Places* as a member of this unit. Age 23, farm laborer, 1860 Floyd Co. Census.

ALTIZER, JESSE R.: 1st SHA. Listed in Amos D. Woods' *Floyd County: A History of Its People and Places* as a member of this unit. Age 35, farmer, 1860 Floyd Co. Census.

ALTIZER, M.: 1st SHA. Listed in Amos D. Woods' *Floyd County: A History of Its People and Places* as a member of this unit.

ANDERSON, JOEL C.: 1st SHA. Listed in Amos D. Woods' *Floyd County: A History of Its People and Places* as a member of this unit. Age 30, farmer, 1860 Floyd Co. Census.

ANDERSON, T. W.: 2nd SHA. Pvt. Recd. clothing 11/12/64. NFR.

ANDERSON, WILLIAM H.: 1st SHA. Pvt. Postwar ref. From Floyd Co., Va. Age 19, farm laborer, 1860 Floyd Co. Census.

ANDERSON, ______: 1st SHA. Pvt. Postwar ref. Killed at Dover, near Aldie.

ANTONIO, C.: 2nd SHA. Pvt. Enl. 7/10/62 in Richmond by Lt. Shepherd. Present 1/64-7/8/64 when captured. Parole given at Burkeville, Va. 4/26/65. Signed with an X.

ARGYLE, E. H.: 1st SHA. Pvt. Paroled at Columbia, Va. by 1st N.Y. Mtd. Rifles 5/4/65.

ARNOLD, FRANK A.: 1st SHA. Pvt. Only record in file shows pay record for 4/30/63. Res. of Hookstown, Baltimore Co., MD.

ARNOLD, JOHN A.: 2nd SHA. Pvt. Occupation: hatter. Enl. w/Co. H, 17th Va. Inf. at Alexandria, 4/17/61, age 19. Absent wounded, 8/31/62. Present 4/43. Shown as AWOL on rolls 11/62- 2/63 (probably the time period he was serving with SHA.) Returned to 17th while at Culpeper CH. Under arrest 10/63. Apparently got out of CM. Captured 4/65 and took Oath at Alexandria. Moved around 1869 to Bloomington, Ill., where he died 10/1/06.

ARNOLD, JULIAN: 2nd SHA. Pvt. Returned to 5th Va. Inf. Killed by thieves at Alexandria.

AUSTIN, WILLIAM R.: 2nd SHA. Pvt. Paroled 4/29/65 in Ashland to return to Louisa County.

AVERY, WILLIAM: 2nd SHA. Pvt. Deserted at Culpeper CH.

BABER, LEONIDAS L.: 1st SHA. Pvt. Enl. 5/27/62, Richmond. Present 11/2/63 - AWOL (1/1/64). Absent with leave (4/64). NFR.

BABER, W. A.: 2nd SHA. Pvt. Clothing recd. 12/28/64.

BAKER, ANDREW JACKSON: 2nd SHA. Pvt. B. 1840. Pvt. Enl. 9/1/62 in Maryland. Blacksmith 11/1/62 - 1/1/63. Rate of pay 40 cents. Present through end. Noted as a blacksmith again in file 5/1/63-10/31/63. Clothes recd. 10/18/64 and 11/12/64. Res. of Mt. Airy, Carroll Co., MD.

BAKER, SAMUEL: 2nd SHA. Pvt. Scant records available. Noted at GH #3 with acute pneumonia 8/63. Shows died 8/15/63. However, other records show died 8/13/64 at GH #3 in Lynchburg. Effects ($5.50) received 8/30/64. Certificate #2367. Yet another final note in his file states POW paroled at Farmville between 4/11 and 4/21/65. J.S. BAKER may have been the one who died. Res. of Royer's Forge, Baltimore Co., MD.

BALCH, ______: 1st SHA. Postwar ref.

BALL, R. J.: 2nd SHA. Pvt. Clothes recd. 11/29/64. NFR.

BARTON, CHARLES M.: 1st SHA. Pvt. Postwar ref. Killed 5/22/62. May have also served in the Jackson Artillery.

BARTON, DAVID R.: 2nd SHA. Pvt. Postwar ref. Killed 12/13/62. May have also served in the Jackson Artillery, and the University Battn. VA Inf.

BASHAM, HILERY H.: 1st SHA. Listed in Amos D. Woods' *Floyd County: A History of Its People and Places* as a member of this unit. Age 29, farmer, 1860 Floyd Co. Census.

BATES, JULIUS E.: 1st SHA. Pvt. B. Arkansas. Roll of POW's captured at Westminster, 7/2/63. Sent to Ft. McHenry 7/5/63. Then to Ft. Delaware 7/7/63. Wishing to enl. in U.S. service 8/30/63. NFR.

BAUGHN, GEORGE DALLAS: 2nd SHA. Pvt. Teamster 9/21/64 to end. Surrendered at Appomattox 4/9/65.

BAUGHN, JAMES C.: 2nd SHA. Pvt. Teamster for the same brother G.D. shows above. Also surrendered at Appomattox 4/9/65.

BEALL, LLOYD: 1st SHA. Pvt. Had served for at least 14 years in U.S. service, 11 of which was with the artillery. Left as U.S. captain. Enl. 8/1/63, Richmond. Absent with leave 8/12/63. At Chimbo #2 with spasmodic asthma 2/20-4/16/64. Deserted from hospital. Transf. to Md. line per S.O. 184/42, 8/5/64. Captured at Strasburg 10/9/64. To Pt. Lookout 10/13/64. Exchanged 2/13/65. Present at Camp Lee 2/20/65. Recommended for a commission by Maj. Breathed 3/20/65; "He has been a private in my command for the last ten months has displayed coolness, gallantry... In my opinion he is in every particular qualified for a commission." Endorsed by M.Gen. Fitz Lee. Apparently at the opening of the war was a mustering officer with the U.S. While in U.S. service left Kansas with intent of resigning. Letter of L. Beall dated 2/15/64 stated "...I frequently refused to do duty in the United States Service, especially after the first battle of Manassas at which time I said that I would never again take up arms against the South for which I was placed under arrest and was not released until I consented to do the duty of "mustering officer""Since his wife made it impossible to travel he was "compelled to perform that duty." Stuart's endorsement went all the way back to 12/15/63.

BEARD, ROBERT (HOOPER): 1st SHA. Pvt. B. S.C. Enl. 1/7/62, Talledega, Alabama. Record states deserted near Brandy Station 6/6/63 "No Interest South." Captured in the def. Of the South Potomac 6/13/63 and sent to OCP. Oath taken. Sent to Philadelphia.

BENNETT, LIVINGSTON ORIC: 1st SHA. Pvt. Enl. 10/11/__ in Martinsburg. Present until captured on the Potomac 3/4/63 and sent to Old Cap. Prison. Apparently exchanged as shows present by 8/10/63 when present in Chimbo. #1 until 8/19/63. Transfer orders to Md. line per SO #100/2; revoked per SO #177/7 7/28/64. Took oath in Danville, Va. 4/28/65; destination his residence at 229 N. Eutaw St. in Baltimore, Md. At Soldier's Home in Pikeville, Md. in 1894 (1335 York Rd. listed as residence.)

BENNETT, WILLIAM V.: 1st SHA. Pvt. Resident of Herford Co., Md. Occupation: farmer. On rolls from 1/64 until captured at Strasburg 10/9/64. Sent to Pt. Lookout and released on oath 5/12/65. Was still due his bounty when captured.

BERRY, WALLER or WALLACE: 2nd SHA. Pvt. Enl. 3/7 or 27/62 at Culpeper. Pres. through end. Special notes state Wagon Master 7/9/63 - 10/31/63 at a rate of pay of 25 cents. Also riding private horse on detached service (11/3/64) through 1/20/65. Postwar note stated that he was assigned with the Bttn. ordnance wagons.

BEST, JAMES ELIAS: 2nd SHA. Pvt. B.7/25/45, Rappahannock Co. D. 5/24/35, Rappahannock Co.

BINI, JOHN: 2nd SHA. Pvt. Resident of New Orleans, La. Age 43, occupation: laborer. Enl. 2/14/63 in Richmond. Wounded 5/2/63, operated on by surgeon Randolph on 5/4, ball through lower third of thigh. Leg amputated at middle third on 5/4. Admitted to hospital 5/19, general health was good and stump nearly healed; 60 day furlough 6/11/63 at residence in Richmond. At Wayside Hosp. or G.H. #9 in Richmond 9/30/63. Furlough 10/1/63 for 60 days. At Farmville G.H. 5/27/64 (clothing recd.) Transfer from Farmville to Charlottesville 6/3/64. At Louisiana hospital 10/21/64 (clothing recd.). Ordered to Invalid Corps P.A.C.S. 10/21/64, Mobile, Alabama. NFR.

BIRD, EDWARD T.: 2nd SHA. Pvt. Enl. 3/15/62 at Franklin by Capt. Hale. Pres. through 1/20/65 with private horse. NFR.

BIRD, WILLIAM L.: 1st SHA. Listed in Amos D. Woods' *Floyd County: A History of Its People and Places* as a member of this unit. Age 27, farmer, 1860 Floyd Co. Census.

BLASSINGAME, ____: 2nd SHA. Pvt. Postwar ref.

BLOUNT, S. M.: 1st SHA. Pvt. Enl. 8/7/__ in Richmond by Capt. Johnston. Present (8/31/64). NFR.

BOLINGER, WILLIAM: 2nd SHA. Pvt. Enl. 2/1/62 in Alabama. Present until on furlough (1/20/65).

BOLLMAN, JOHN M.: 1st SHA. Pvt. Res. Of Baltimore. Enl. 10/1/62 at Martinsburg by Capt. Breathed. Present until at GH #9 at Wayside 6/1/64. Postwar records state wounded "blown up on caisson" at Union, Va. 11/2/62. Transferred to Co. K, 1st Va. Cav. 8/14/64. Transferred to Co. K, 1st Md. Cav. 8/15/64. Res. of Baltimore. D. by 1903.

BOONE, JACOB: 1st SHA. Listed in Amos D. Woods' *Floyd County: A History of Its People and Places* as a member of this unit. Age 32, farmer, 1860 Floyd Co. Census.

BOOTHE, HENDERSON: 2nd SHA. Pvt. Enl. 4/15/62 in Richmond. Pres. through 1/20/65.

BOOTHE, J. W.: 1st SHA. Listed in Amos D. Woods' *Floyd County: A History of Its People and Places* as a member of this unit.

BOOTHE, LEWIS: 1st SHA. Listed in Amos D. Woods' *Floyd County: A History of Its People and Places* as a member of this unit. Age 30, farmer, 1860 Floyd Co. Census. May have also served in the 36th VA Inf.

BOOTHE, NOAH R.: 1st SHA. Listed in Amos D. Woods' *Floyd County: A History of Its People and Places* as a member of this unit. Age 20, farmer, 1860 Floyd Co. Census.

BOOTHE, THOMAS: 2nd SHA. Pvt. Enl. 4/15/62 in Richmond. Present through 1/20/65. NFR. Listed in Amos D. Woods' *Floyd County: A History of Its People and Places* as a member of this unit.

BOURKE, HUGH: 1st SHA. Pvt. Violated 9th Article of War. NFR.

BOWLES, WILLIAM J.: 2nd SHA. Pvt. Clothing recd. 12/28/64. NFR. May have also served in the 10th VA Cav.

BOYD, ____ : 1st SHA. Pvt. Postwar ref. From Baltimore.

BOYDTON, WILLIAM L.: SHA. Postwar ref. Notes include something about misbehavior before the enemy. See also Shoemaker's Lynchburg Artillery.

BRANCH, CHARLES: 1st SHA. Pvt. Enl. 6/1/63 at Richmond by Lt. Johnston. Wounded and at Jackson Hospital 6/2/64 (Horse wounded or killed as well). Suffered shell wound to arm. To Cpl. 7/17/64. NFR.

BRANCH, WILLIAM: 1st SHA. Pvt. Postwar ref.

BREATHED, F.[B] H. G.: 1st SHA. Pvt. Buried at Sharon Cem., Middleburg, Va.

BREATHED, JAMES W.: 1st SHA. 1st Lt.-Capt. B. 2/13/38 at Funkstown, Md. son of John W. and Ana MacGill Williams Breathed of "Boyuca" near St. James College in Hagerstown, Md. Educated at Military School in Urbanna (school run by John Robert Jones, later general A.N.V.). Entered St. James College. Later attended Maryland University and Medical College. Graduated 1858 with degree in medicine. Went west to practice. Practicing in St. Joseph, Mo. in 1861. On return trip east rode in train with J.E.B. Stuart. Enl. 4/19/61 at Martinsburg in Co. B, 1st Va. Cav. Appointed Lt. of cav. 11/18/61. Transferred to SHA 11/20/61. 1st Lt. 3/23/62. Captain 9/22/62 to rank from 8/9. Attempted to resign 9/25/63 but strongly disapproved by Generals W. C. Wickham, Fitz Lee and Stuart. Major 2/27/64. Commanded Horse Arty. Bttn. until slightly wounded 5/10/64 at Yellow Tavern. Wounded again 6/24. Detached with Gen Early 8/64. Paroled at Winchester 4/24/65. 5'11", light complexion, dark hair, blue eyes. After war established medical practice in Hancock, Md. D. 2/14/70, age 32. Buried at St. Thomas Episcopal Church, Hancock, Md. George Wilmer Brown annotated in a postwar note that Breathed's middle initial was "N."

BRIEUX or BREAUX, EDMOND: 2nd SHA. Pvt. Enl. 4/1/63, Culpeper. Deserted (1/20/65). 3rd New York Cavalry picked-up and forwarded him from Bernard's Mills via Portsmouth.

BROOKE, ROBERT M.: 2nd SHA. Pvt. Enl. ? At Chimbo #3 with scabies 8/16 - 8/29/64. Transferred to Thornton's Battery per S.O. #38/39, 2/15/65. Occupation: cattle driver. Lived after the war (1894) at 416 N. Carrollton Ave., Baltimore. In Md. Soldier's Home also in 1894.

BROWN, DANIEL: 2nd SHA. Pvt. Listed only as wagoner from 11/1/62 - 1/1/63. NFR. Listed in Amos D. Woods' *Floyd County: A History of Its People and Places* as a member of this unit. Age 24, farmer, 1860 Floyd Co. Census. Also appears to have served in the 21st VA Cav.

BROWN, GEORGE WILMER: 2nd SHA. Captain. B. son of Rev. Richard Templeton Brown (rector of the church at Fairfax CH, a direct descendant of Edward Brown, who came to Va. in 1607. E. Brown's wife was Margaret Templeton. G.W. Brown's mother was Marion Rebecca Wilmer (m. to R. T. Brown in Alex., Va. 1837) whose ancestor had come to Maryland in 1660. Grandfather, William Holland Wilmer was President of the College of William and Mary and was buried in the churchyard of the Bruton Parish Church, Williamsburg. Entered VMI 7/59. Went to Richmond with the Corps of Cadets in 1861 and was drill master for Co. C (Patrick Henry Rifles), 15th Virginia Infantry. Joined Pelham's Battery as a private. 2nd Lt. Captain, 2/18/65. Captured at Appomattox. After the war resided mostly in western N.Y. where he engaged in real estate business at 406 Presidential building, Buffalo, N.Y., Lived with his niece, Miss Marian Van Horn, in Bancroft, Va. some years before his death. D. in Johns Hopkins University hospital. Services in Epiphany church before being interred in Arlington National Cemetery.

BROWN, GEORGE W.: 1st SHA. Listed in Amos D. Woods' *Floyd County: A History of Its People and Places* as a member of this unit. KIA during the Seven Days' Battles around Richmond in summer 1862. Age 31, Farmer, 1860 Floyd Co. Census.

BROWN, JAMES F.: 1st SHA. Cpl. B. 7/9/41 in Jefferson Co., Va. Moved with his family to Washington Co., Md. in early childhood. Educated at St. Mary's College. Enl. Co. K (2nd), 1st Va. Cav. in 1861. Served as civilian courier for Gen. J.E.B. Stuart. Transferred to SHA. Captured at Hagerstown, Md. 7/7/63. Sent to Ft. Delaware. Transf. to Pt. Lookout. Exch. 2/10/65. Surrendered at Appomattox CH 4/9/65. Lived at Hagerstown, Washington Co., MD postwar. D. from Bright's Disease 12/27/98, age 57, buried in Loudon Park, Baltimore. Survived by wife and one daughter.

BROWN, JAMES TEMPLETON "Temp": 2nd SHA. Cpl. Enl. 12/24/62, Richmond. Present through 1/20/65 when listed as 3rd cpl.

BROWN, JOHN: 1st SHA. Pvt. Enl. 5/15/__ in Richmond by Lt. Col. Pate. Present until at CSA GH Charlottesville with pneumonia 3/11/64. Died 3/13/64. Effects recd. 4/16/64, cert. #842. $41.25 claim #4449. Res. of Calvert Co., MD.

BROWN, JOHN: 2nd SHA. Postwar records.

BROWN, T. J.: 2nd SHA. Sgt. Died near Pohick in 1863 after having taken sick and being left.

BULGER, JOHN: 1st SHA. Postwar roster states "Mississippi."

BUNCH, RICHARD: SHA. Postwar reference. From Tennessee.

BURFORD, WILLIAM E.: SHA. Postwar reference. AWOL. See also Shoemaker's Lynchburg Art.

BURGER, THOMAS: 1st SHA. Pvt. Enl. 7/24/62. Sentenced to "buck and gag" 5/25/63. To be bucked 2 hours at a time with 2 hour intervals 5/27/63 then "log drill" by Major Beckham 5/28/63. Captured at Westminster 7/2/63. To Ft. McHenry then Ft. Columbus, N.Y. Harbor, then Pt. Lookout, then Elmira 10/14/64. For exchange 10/29/64, apparently did not occur. Paroled 2/18/65 at Pt. Lookout. NFR.

BURGESS, THOMAS: 1st SHA. PWR only.

BURGESS, JOHN L.: 1st SHA. Listed in Amos D. Woods' *Floyd County: A History of Its People and Places* as a member of this unit. Noted that he was taken POW at Woodstock on 9/23/64, sent to Point Lookout, Md., where held until paroled on 5/12/65. Age 27, farmer, 1860 Floyd Co. Census.

BURKE, HUGH: 1st SHA. Pvt. Enl. 7/24/62 in Richmond. CM ordered for viol. of 9th article of war per GO #68, 6/3/63. Confined 5/19/64 near Culpeper CH by Capt. Breathed. Revoked CM per SO#201/24, 8/25/64. Was to transfer to Md. line 4/29/64. NFR.

BURKE, JOHN: 1st SHA. Postwar roster states "Mississippi."

BURKE, MARTIN or RICHARD: 1st SHA. Pvt. Bugler. Res. of Frederick City, Md. Enl. 12/15/62 at Richmond. At CSA GH Charlottesville 8/16/63 -9/16/63. Wounded and captured 10/17/63 battle of Buckland, Fauquier Co. Surgery performed by Surg. William Murray at Lynchburg #1. At CSA GH Charlottesville 4/16/64 - 5/3/64. Transferred to Lynchburg. Postwar account states lost leg at Blackburn's Ford. Applied for wooden leg. Amputated at ankle due to shell wound.

BURWELL, GEORGE H.: 2nd SHA. Pvt. See also 2nd VA Inf.

BURWELL, ROBERT POWEL PAGE: 2nd SHA. 2nd Lt. Son of Nathaniel and Dorothy P. Burwell of "Glenvin," Clarke Co. Member of "Stonewall" Brigade early in war at age 17. After 11 months joined SHA and appointed Lt. after Sharpsburg. Date of rank 9/17/62. Commissioned 1/9/63. Wounded at Brandy Station 8/1/63. D. 8/31/63, aged 19 years, 7 mos. and 14 days. Buried in the Old Chapel, Clarke Co., Va. near the grave of M.W. Henry and John Esten Cooke. See also 2nd VA Inf.

BUSIC, JOHN: 2nd SHA. Pvt. Disabled in hospital (11/3/64). NFR. B. 1840 Wilson District, Grayson Co., VA. Age 21, Farmer, 21, 1860 Grayson County Census. Age 36, Farmer B. VA, 1870 Alleghany Co., NC Census. D. 1919 Piney Creek, Alleghany Co., NC. May have also served in the 23rd VA Inf.

BUSKELL, RICHARD: 1st SHA. Pvt. B. ca. 1837. Student age 23, Love's Mill, Washington Co., 1860 census. Enl. 5/14/61, Abingdon, Va. With Co. D (2nd), 1st Va. Cav. Present until on detached duty 11 - 12/61. Transferred to the 1st SHA, 1/2/62. KIA Fredericksburg, 1863?

BUTLER, _________: 1st SHA. Pvt. Enl. 2/27/63? in Richmond. Absent in hands of enemy 11/2/63.

BUTTERLY, THOMAS: 2nd SHA. Quartermaster Sgt. Enl. 4/30/61, new Orleans. Later transferred to the 2nd SHA. Present through 1/20/65. Residence: Shreveport, La. NFR.

BYRON, _____: 1st SHA. Postwar reference. Killed.

BYRON, REUBEN: 2nd SHA. Bugler. Enl. 1/12/62, Richmond. Captured 6/63 in Maryland. Exchanged and present (1/20/65). NFR.

CAHILL, MARTIN: 1st SHA. Pvt. Postwar roster has him listed as from Mississippi and a sergeant. Enl. 4/61 in Richmond by Capt. Lewis. Enl. again 7/3/62 in Richmond. Captured 6/30/63 at Westminster. To Ft. McHenry, Ft. Delaware 7/8/63. To Ft. Columbus, NY Harbor 9/26/63. Exchanged 2/13/65. Pres. in Camp Lee 2/18/65. See 18th Mississippi Infantry. D. 4/25/97 in Fredericksburg, Va., age 65. See Fredericksburg *Free Lance* 4/27/97, p.3, c2 for obituary.

CALBRETH, JOHN: 1st SHA. Pvt. Enl. ? WIA 11/2/62, Union. Transferred to Co. K, 1st Va. Cav., 8/14/64. Transferred to Co. K, 1st Md. Cav., 8/15/64. NFR.

CALLAHAN, A. M.: SHA. Postwar reference records "25 days leave of absence."

CARDWELL, DAVID: "Davy" 2nd SHA. Pvt. Enl. 7/22/63, Richmond. Occupation listed as clerk. Present through (1/20/65). Paroled 5/5/65, Ashland. Alive in 1913 in Richmond. Later, Asst. General Freight Agent of the Southern RR, Columbia S.C.

CARNEY, M. J.: 1st SHA. Pvt. Deserter 62-63. NFR. See also 10th VA Cav.

CARR, JOHN: 2nd SHA. Pvt. Enl. 4/15/62, Richmond. Listed with his private horse. Present through 11/64. NFR. Listed in Amos D. Woods' *Floyd County: A History of Its People and Places* as a member of this unit. Age 28, farmer, 1860 Floyd Co. Census.

CARR, JOSEPH: 1st SHA. Listed in Amos D. Woods' *Floyd County: A History of Its People and Places* as a member of this unit. Age 16, farm laborer, 1860 Floyd Co. Census.

CARRICO, WILLIAM HENRY: 2nd SHA. Pvt. Enl. 3/27/62, Culpeper. Listed with private horse. Present through (11/12/64). NFR. Age 24, farm laborer, 1870 Grayson Co., VA Census.

CARRINGTON, _________: 2nd SHA. Postwar reference.

CHAMBERLAINE, RICHARD: 2nd SHA. Pvt. B. 1843, Norfolk, Va. Parents were R.H. and _____ Wilson Chamberlaine. Matriculated VMI in 1860. Grad. 7/4/63. Enl. 3/27/64, Gordonsville. At Jackson Hospital with dysentery, 6/9/64. Fell from horse and listed at Chimbo # 3, 8/16/64. Severely burned by discharge of cannon at Five Forks. Paroled 4/21/65, Ashland. After the war engaged in business in New York and Norfolk. D. 11/30/79, age 36 in N.Y. Never married.

CHAPMAN, N.: 1st SHA. Pvt. Enl. 9/23/63? in Richmond by Capt. Hawkins. Pres. through 8/31/64. NFR.

CHARLES, ____: 1st SHA. Postwar ref.: Violated 23rd Art. of War. Resided after war in St. Louis.

CHICHESTER, ASHTON M.: 2nd SHA. Sgt. Enl. 3/10/62, Culpeper. Wounded and at Charlottesville GH 7/9/63. Returned 7/25/63. Owned private horse. Noted as promoted for valor and skill 8/29/64 by McGregor. 4th

Sgt. (1/20/65). NFR. Postwar resident of Georgetown, D.C. D. ca.4/14/16 (date of Richmond Newspaper obit.)

CHICHESTER, DANIEL McCARTY: 2nd SHA. Battalion Sgt. Major. Enl. 5/17/61, Fairfax. Transferred to the ranks of the SHA. Private horse. Detached with "other portion of the battalion" (7/24/64). Present through (1/20/65). NFR. See also the 17th VA Inf.

CLARKE, WILLIAM G.: 1st SHA. Postwar ref.: Viol. 9th art. of war.

CLATTERBUCK, H.: 1st SHA. Pvt. Enl. 2/1/? at Madison C.H. by Capt. Strother. Present 4/30/64 - 8/31/64. Pay due for private horse 9 months, 15 days, $114.

CLEM, WILLIAM B.: 1st SHA. Pvt. Enl. 2/12/62 at Martinsburg by Capt. Koontz. Absent from 5/2 to 8/5 without leave. Paroled at New Market 4/24/65; age 28, 5'7", fair complexion, lt. hair, hazel eyes. See also 7th Va. Cav.

COCHRAN, J. H.: 2nd SHA. Pvt. Clothing issue listed 11/1/64. NFR.

COIT, N.: 1st SHA. Pvt. Enl. in Asheville, Alabama by Lt. Bradford 12/15/61. Enl. again 12/25/ in Richmond. At Chimbo. #2 with debility 3/8/62-5/3/62. Attached to hospital as nurse 5/9/62. At hospital wounded 7/27/63. Admitted to Wayside GH #9 1/17/64; to duty 2/18/64. Pres. 4/1/64. Captured at Strasburg 10/9/64. Released from Pt. Lookout 5/12/65.

COLLIER, EDWARD J.: 1st SHA. Pvt. Enl. 7/25/__ at Richmond. AWOL 11/2/63. NFR. See also 1st VA Inf.

CONNOR, ALBERT: 1st SHA. Listed in Amos D. Woods' *Floyd County: A History of Its People and Places* as a member of this unit. Age 26, farmer, 1860 Floyd Co. Census.

CONNOR, ANDREW: 1st SHA. Pvt. Enl. 3/27/__ at Gordonsville by Capt. Breathed. Present as blacksmith 4/63. Wounded at hospital (GSW) 5/64. D. 5/21/64 at Pratt hospital, Lynchburg. Age 14, 1860 Floyd Co. Census.

CONNOR, DANIEL R.: 1st SHA. Listed in Amos D. Woods' *Floyd County: A History of Its People and Places* as a member of this unit. Age 45 or age 32, farmer, 1860 Floyd Co. Census.

CONNOR, JONATHAN, Sr.: 1st SHA. "captain of the forge" B. 4/20/31 in Floyd Co. Age 29, farmer, 1860 Floyd Co. Census. M. Mahelea Thrash 7/27/52. Occupation: farmer/blacksmith. Artificier. Enl. 4/16/62 in Richmond by Pelham. Served to the end with his own horse. Had two or three slaves who accompanied him on several campaigns. Supposedly when Gen. R.E. Lee came into the camp one day, he asked if he could have Traveler shoed. Connor replied that he could have it done in five minutes, Lee could not believe it, but it was done under that time. D. 4/18/14 in Floyd Co. Listed in Amos D. Woods' *Floyd County: A History of Its People and Places* as a member of this unit.

CONNOR, WILLIAM P.: 1st SHA. Pvt. Enl. 5/16/__ at Hanover CH. AWOL 10/9/64. NFR. Listed in Amos D. Wo-ods' *Floyd County: A History of Its People and Places* as a member of this unit. Age 36, farmer, 1860 Floyd Co. Census.

CONROY, DENNIS: 1st SHA. Pvt. An original member of the Newtown Artillery. Enl. 7/1/61, Newtown. Present through 10/61. Orderly for Maj. Pelham. Enl. 11/5/62 at Camp Alexander by Capt. Pelham. Present through 10/64. Bay horse KIA 10/9/64 at Tom's Brook; $3,000.00. Horse was mustered into service 9/1/64. Paroled at Greensboro, NC with J. E. Johnston's Army 4/26/65.

CONSTANTINI, ANTONIO: 2nd SHA. Postwar ref.

COODE, DEMETRIUS A. G.: 1st SHA. Cpl. Enl. 7/11/62 in Richmond by Lt. Shepherd. Postwar roster states he was wounded at White House, no date listed. Pres. until wounded and sent to hosp. 6/15/64. GSW in hand. Transfer top Md. Line revoked. Horse killed at Yellow Tavern 5/12/64; $1566.66. NFR. Also served in the 43rd Battn. VA Cav. (Mosby's). Res. of Skipping Point, St. Mary's Co., Md.

COOK, J. D.: 2nd SHA. Sgt. Enl. 4/20/61, New Orleans. Transferred to the SHA. Pres. Through (11/12/64). NFR.

COOK, JOHN FRANK: 2nd SHA. 2nd Sgt. B. 1840, Virginia. Laborer, Emory, Washington Co., 1860 census. Enl. 5/14/61, Abingdon in Co. D, (2nd), 1st Va. Cav. Detached to SHA 11 - 12/61. Transferred to SHA, 1/2/62. Record shows him as enlisting again 12/11/61, Centreville. Absent on horse detail (7/24/64). Abs. with leave (8/9/64). Owned private horse. Present through 11/64. WIA? Postwar resident of Socorro, N.M., 1914.

COON, JOHN: 2nd SHA. Pvt. Enl. 8/1/62, Richmond. Present through 11/64. NFR. See also 1st VA Cav.

CORBIN, _______: 2nd SHA. Postwar reference states killed at Chancellorsville.

CORNEAN, PHILLIP: 2nd SHA. Pvt. Postwar ref.: Killed on Charles City Rd. below Richmond, 1864.

CORNEAU, ____: 2nd SHA. Postwar ref.

CORRINGTON, WILLIAM: 1st SHA. Pvt. Postwar ref.: Wounded near Winchester.

COSGROVE, T. WILLIAM: 1st SHA. Lt. Enl. 7/4/62 in Richmond by Lt. Shepherd. 3rd Sgt 11/63. 2nd Lt. 2/27/64. Commissioned 5/9/64, to date 2/27/64. Lt. Transferred to Md. Line per SO #100/2, 4/29/64. Postwar resident of Park Mills, Frederick Co., Md. D. at Mt. Ephrain, Montgomery, Md.

COSTIGAN, COSGROVE D.: 1st SHA. Pvt. Postwar ref.: Killed at Union, Va. 11/2/62. Also served in Edelin's Maryland Heavy Artillery. Res. of Oakville, St. Mary's Co., Md.

COVINGTON, GEORGE WASHINGTON: 1st SHA. Pvt. Listed in postwar roster as being from North Carolina. Enl. 2/16/62 in Richmond by Lt. Johnston. Absent sick with asthma first 11/2/63 at Lynchburg GH #1, then 12/20/63 at Moore Hosp. Furlough 1/4/64. Wounded 8/21/64 at Berryville and died 9/13/64. Buried in Winchester Cemetery #1305.

CRABBE, ROBERT A.: 1st SHA. Pvt. From Westmoreland Co., Va. Enl. 3/28/6?. Captured at Strasburg 10/9/64. From Harper's Ferry sent to Pt. Lookout where exchanged from on 3/28/65. See also 15th VA Cav.

CRANDELL, CHARLES P.: 1st SHA. Pvt. Enl. 7/1/62 in Richmond by Lt. Johnston. Left sick in enemy lines and captured at Kearneysville 7/17/63. "Claimed to be a Lt." To Johnson's Island, Ohio and to Pt. Lookout, Md. for exchange 3/21/65 Released on oath 5/21/65. Served as Lt. in the 19th Battn. VA Heavy Art.

CRANDELL, GEORGE P.: 1st SHA. Pvt. Postwar ref. Member of UCV group in Washington, D.C. D. by 1910.

CROOK, OLIVER T.: 2nd SHA. Pvt. Enl. 9/2/62, Richmond. Captured 11/64. NFR. Postwar records shows him as deserted. See also 43rd Battn. VA Cav. (Mosby's).

CROSS, CHARLES N.: 2nd SHA. Pvt. Enl. 9/1/62, Richmond. Captured at Culpeper C.H. Released on parole 10/6/63. There was a note in file regarding his stay at OCP; "father is in miserable health & mother beseeches that he may be allowed to go to Alexandria." Returned to duty 10/15/63. Listed as age 19 (9/23/63). Note states released on oath 10/28/63. Captured again ? 11/64. Surrendered 4/13/65; residence Maryland. Destination: Philadelphia. Postwar reference shows him as deserted.

CROYEAN, EDWARD: 2nd SHA. Cpl. - Lt. Enl. 9/25/62, White Post. 3rd Cpl. (7/24/64). 2nd Lt. 6/1/64. Surrendered at Danville, 5/12/65. Took oath; residence listed as Baltimore with the same place as destination. D. Baltimore, 1870.

CULBREATH, JOHN: 1st SHA. Pvt. Enl. Richmond 7/13/62. Postwar record states he was wounded "blown up" at Union, Va. 11/2/62. Transferred to Co. K, 1st Va. Cav. 8/14/64. Transferred to 1st Md. Cav. 8/15/64. Postwar resident of Baltimore, Md. D. 2/20/12, age 74. Buried in Loudon Cemetery, Baltimore.

CUNNINGHAM, T. H.: 2nd SHA. Pvt. Enl. 6/1/64, Hanover. Present through (1/20/65). NFR.

DABNEY, WALTER S.: 2nd SHA. Bttn. Ord. Sgt. - Lt. Enl. 3/1/62, Centreville. Lists enlisting officer as J. E. B. Stuart. Private horse. Listed as Lt. and POW when took oath 5/15/64. Residence: Louisa Co.

DAVIS, F. T.: 2nd SHA. Pvt. Enl. 6/8/61, Campbell. Private horse. Wounded and at CSA GH, Richmond, 6/4/64. Transferred to Chimbo., 7/25/64.Chronic dysentery at Chimbo #2, 7/25/64 - 8/4/64. NFR.

DAVIS, FLETCHER A.: 2nd SHA. Pvt. Enl. 3/1/62, Richmond. At Chimbo. #9, 5/3/63. Captured ? But escaped 8/64. Present through (1/20/65). NFR. Postwar record stated that he left SHA at Culpeper CH, 1863 returned at Petersburg in 1864.

DAVIS, JOHN: 2nd SHA. Pvt. Enl. 3/27/62, Culpeper. Private horse. AWOL (8/9/64). Present (11/3/64). Present through (12/28/64). NFR.

DeMAINE, JOHN: 2nd SHA. Pvt. Enl.? Captured at CCH 9/14/63. Sent to Old Capitol Prison 9/5/63? Took oath 11/14/63. Postwar records state he left SHA at Culpeper CH, 1863, returned while in Petersburg, 1864. Present through (1/20/65). NFR. Joined UCV in Alexandria, 11/92.

DILLARD, A. S.: 1st SHA. Pvt. Enl. 5/29/__ in Richmond by Lt. Johnston. Absent on detached service 1&2/64. Detailed with the Nitre and Mining Bureau, 3rd Regt of Local Def. Troops, Richmond per SO#17/24, 1/21/64. NFR.

DILLER, ANTHONY JOHN: 1st SHA. Pvt. Enl. 5/29 or 6/12/62 in Richmond by Lt. Johnston. At GH #21 Richmond (8/63) Wounded. Postwar roster states he was wounded at Smithfield, Va. At hospitals through 11/11/63. Absent on detached service 1&2/64., per SO#16/3, and again under SO#17/24 1/21/64. Paroled at Danville 4/28/65. NFR.

DINWIDDIE, MARSHALL: 2nd SHA. Pvt. B.1845. Enl. 11/1/63, Orange CH. Private horse. Present through (1/20/65). Paroled at Louisa CH, 5/13/65. Member of J.B. Strange UCV Camp. D. 5/17/16.

DINWIDDIE, WALTHALL: 2nd SHA. Pvt. B.1847. Only record of clothing issue 11/64. Paroled at Louisa CH, 5/13/65. D. 1909.

DODSON, R. TOWN: 1st SHA. Pvt. Enl. 1/7/62 at Manassas by Capt. Pelham. Furlough by SO#104/7, 4/29/63. Paroled 5/12/65, Farmville, Va.; former res. Washington, D.C., destination the same. Postwar roster has him listed as sergeant major.

DOGGETT, DAVID: 2nd SHA. Pvt. Enl. 11/10/63, OCH. Pres. Through (11/3/64). NFR. Postwar record shows transferred to 4th Va. Cav.

DOGGETT, JAMES T.: 2nd SHA. Pvt. Enl. 3/27/62, Culpeper. Private horse. Present through (1/20/65). NFR.

DOMMICO, R.: 2nd SHA. Pvt. Enl. 6/20/62, Richmond. Deserted 5/20/64. NFR.

DONALD, JOHN A.: 2nd SHA. Pvt. Enl. 6/8/61, Campbell. KIA 5/12/64.

DORSEY, JOHN: 1st SHA. Pvt. Enl. 6/30/62. Also of Madison Lt. Arty. of La. Postwar account states he was from Maryland and was wounded near Warrenton, Va. D. from wound.

DOWNING, GEORGE T.: SHA. Postwar ref.: D. Fredericksburg 6/30/12, age 65. No Service Record.

DUDLEY, THOMAS CLIFFORD: 1st SHA. Pvt. B. 2/26/46 at Benvenue, King and Queen County, Va. eldest son of Alexander and Martha Ellen Jackson Dudley. Early education as the pupil of Colonel J.C.Council at Aberdeen Academy in King and Queen County. Joined the militia at the outbreak of the war and accompanied them to Gloucester Point for a tour without actually being mustered into service. Entered VMI in the summer or early autumn of 1861. However, within the year he left and joined the 1st Stuart H.A. (7/26/62). Pelham wrote of Dudley: "In two days of fighting at Groveton Heights and at Sharpsburg, South Mountain and various heavy skirmishes he was conspicuously gallant. He is attentive...eminently qualified to command and discipline a company." Transferred to Co. E, 5th Va. Cavalry. Promoted for meritorious service to 2nd Lt. P.A.C.S. and assigned to duty as enrolling officer in Greensville County. By S.O. 3/1/64 he was relieved from duty as enrolling officer and assigned similar duty in Giles County, continued to

5/10/64. When Federals occupied Dublin, Va. he was forced to leave his post and returned to his home county and procured a horse to ride with Co. E, 5th Va. Cavalry once again as a private. On 6/11/64 at Trevillian Station. Wounded in the leg and taken to the house of Mr. Sumner where the leg was amputated. D. 7/9/64.

DUFFY, BENJAMIN B.: 2nd SHA. Pvt. Enl. 6/2/64, Gordonsville. Detailed at Field Park (11/3/64 - 1/20/65). NFR.

DULANEY, WILLIAM: 1st SHA. Listed in Amos D. Woods' *Floyd County: A History of Its People and Places* as a member of this unit. Age 25, farmer, 1860 Floyd Co. Census.

DULANEY, WYATT: 1st SHA. Listed in Amos D. Woods' *Floyd County: A History of Its People and Places* as a member of this unit. Age 21, farmer, 1860 Floyd Co. Census.

DUMM, JAMES: 1st SHA. Pvt. From Maryland. Enl. 1/11/63 in Richmond by Lt. Johnston. At Chimbo. #1 11-12/63. Absent POW 2/64. AWOL 4/1/64. NFR. Resided in Missouri after the war.

DUNCAN, GEORGE R.: 1st SHA. Listed in Amos D. Woods' *Floyd County: A History of Its People and Places* as a member of this unit.

DURRAND, JOHN: 2nd SHA. Pvt. Enl. 8/1/62, Richmond. Captured at Gettysburg, 7/4/63. Mustered into U.S. service 7/27/63, joined Captain Ahl's Battery.

DUSENBERRY, H. BOWIE: 1st SHA. Pvt. From Maryland. Enl. 6/22/62 in Richmond by Lt. Shepherd. Pres. until at Chimbo. #2 6/64. Deserted 6/16/64. NFR. Res of Annapolis, Anne Arundel Co., Md.

DUVAL, J. E.: 2nd SHA. Postwar ref.: Killed at Chancellorsville.

EAGLES, R. W.: 1st SHA. Pvt. Enl. 3/12/__ at Buckingham by Capt. McKenny. Present until at Chimbo. #5 with scabies 8/7-8/20/64. NFR.

EASTON, RUSSELL S.: 1st SHA. Pvt. At Staunton GH 11/64-12/64. Paroled at Winchester 4/30/65, age 25, 5'8", dark compl., black hair and eyes.

ELAM, THOMAS H. E.: 1st SHA. Pvt. Enl. 7/8/64 at Ream's Station by Capt. Johnston. Captured at Mt. Jackson 3/7/65. Paroled at Ft. Delaware 3/65 Residence: Richmond, blue eyes, lt. hair and complexion, 5'3½".

ELAM, WILLIAM A.: 1st SHA. Pvt. Enl. 5/25/63 in Richmond by Lt. Johnston. Present until captured 10/9/64. To Pt. Lookout and exchanged 1/17/65. At Camp Lee 1/20/65. Paroled at Burkesville, Va. 4/65.

ELMORE, W. T.: 2nd SHA. Pvt. Clothing receipt record only (12/28/64). NFR.

ELSTON, WILLIAM C.: 2nd SHA. Pvt.- 1st Lt. Enl. 2/14/63, Essex Co. Commissioned 9/22/62, date of rank 8/10/62. Postwar record shows resignation of 1/13/63. Supposed surrendered at Appomattox, however, also received parole from Winchester 5/8/65. Resident of Oxford Co., Alabama.

EPPERLY, ADDISON: 1st SHA. Pvt. From Floyd Co., Va. Enl. 4/16/62. Present until temp. detached 1 & 2/64. Pres. 4/64. NFR. Listed in Amos D. Woods' *Floyd County: A History of Its People and Places* as a member of this unit. Age 27, farmer, 1860 Floyd Co. Census.

EPPERLY, PHILIP S.: 1st SHA. Listed in Amos D. Woods' *Floyd County: A History of Its People and Places* as a member of this unit. Age 39, farmer, 1860 Floyd Co. Census. Also Served in the 4th Virginia Reserves.

EPPERLY, WILLIAM H.: 1st SHA. Pvt. From Floyd Co., Va. Enl. 4/16/62 in Richmond Present until temp. detached 1 & 2/64. Present 4/1/64. Listed in Amos D. Woods' *Floyd County: A History of Its People and Places* as a member of this unit. Several possibilities in the 1860 Floyd Co. Census.

EVANS, C. BENTON: 1st SHA. Pvt. Enl. Darkesville, 9/25/62 by Capt. Breathed. Present through 4/64. Postwar roster states he was "Killed in the Valley." NFR.

EVANS, CHARLES A.: 1st SHA. Pvt. B. Amelia Co., Va. Enl. 6/18/62 in Richmond. Present until severely wounded 8/29/62; sent to GH #13 and right arm amputated at shoulder joint. GPO Med-Surg., V. II, pt. II, p.651. Lost arm during war. Most of remaining service done in hospitals. Furlough 6/25/63. Paroled 3/18/65; age 22, 5'?, fair compl., blue eyes, dark hair, occupation clerk. Listed in Maryland Soldier's Home. D. by 1894.

EVANS, SAMUEL TAYLOR: 2nd SHA. Pvt. B. in Virginia. Enl. 4/5/62, Richmond. Seriously wounded at Fredericksburg, 12/13/62. Age 15, merchant. "Bombshell contusion of all internal organs over stomach. Greatly prostrated on admittance - Afterwards daily improving." Transferred from GH #12, Richmond to private home. Detailed by General R.E. Lee 8/27/63 per S.O. #213/6. Detailed to Army HQ and present there through (1/20/65). Courier to Lee's HQ with the mail from A & I Genl. Office. M. 1/9/77 in Obion Co. to Sue A. _____. (B. 8/1844, Blount Co.). Enl. 1861. Wd. at Fredericksburg. Postwar doctor. D. 1/9/90 in Obion Co., Tennessee (Union City). Wife resided in Texas since Spring of 1899.

EVANS, WILLIAM: 1st SHA. Postwar ref.:killed at Chancellorsville.

EWING,___: 2nd SHA. Postwar reference states he was with the Light Brigade at Balaklava.

FAULCONER, JOHN THOMAS: 2nd SHA. Pvt. Postwar ref. B. 1832. D. 1913. See also 13th Va. Inf.

FAUNTLEROY, THOMAS KINLOCH "Flintlock": 1st SHA. Pvt. B. White Post, Clarke Co., 1837. Enl. 6/4/61, Shepherdstown in Co. D (1st), 1st Va. Cav. Transferred to Co. D, 6th Va. Cav. 9/61. Transferred to SHA, 10/61. Promoted 1st lieutenant of artillery. Assigned as ordnance officer, Gordonsville. Assigned to 1st C.S. Light Artillery in La., 1863. Promoted to captain and on staff of Gen. Henry T. Allen in Trans-Miss. Dept. 9/15/63. Paroled at Shreveport, La. 6/14/65. Postwar res. of White Post. Methodist Minister, Rushton, La. in postwar also.

FAY, JOHN: 1st SHA. Cpl. Enl. 7/11/__ in Richmond by Lt. Shepherd. Present until deserted 3/11/65. Took oath and sent to Washington, D.C. 3/18/65.

FAYOLLE, LEWIS: 2nd SHA. Pvt. Enl. 2/62, New Orleans. Transferred to the SHA. Pres. through (1/20/65). Paroled at Winchester 4/30/65.

FEBRY, MOSES A.: 1st SHA. 2nd Lt. Corporal when sent 7/16/62 to Floyd Co. With three other men to round up deserters. Commissioned 9/22/62, date of rank 8/10/62. Resigned 1/63. Captured at Middleburg 8/63. Rec'd. pension while residing in Alleghany, Md. Listed in Maryland Soldier's Home. D. 8/12/92, buried in Loudon Park Cemetery, Baltimore.

FENNEY, L.: 1st SHA. Same as Finney, Leigh, 2nd SHA? Pvt. Enl. 4/23/61, Richmond. Transferred to the 1st SHA. Sick in hospital at Chimbo. # 3 & 4, 10/6 - 12/10/64. NFR.

FERRARI, JOSEPH: 2nd SHA. Pvt. Enl. 4/5/63. Present through (1/20/65).

FIELD, GEORGE: 2nd SHA. Pvt. (See 17th Va. Inf.) Transferred to the SHA. Captured CCH 9/16/63. Sent to OCP. From OCP sent to Pt. Lookout, 10/27/63. Exchanged from Pt. Lookout 2/24/65. Occupation carpenter.

FINKS, JOHN: 2nd SHA. Pvt. Enl. 3/27/62. At home disabled (11/3/64). Absent wounded (1/20/65). NFR.

FINNEY, LEIGH: 2nd SHA. Pvt. Enl. ? At GH #9, Richmond 5/16/64 - transferred to Chimbo #4, 5/17. In private quarters 5/19. Returned to duty 10/6/64. Paroled at Manchester 4/21/65.

FISHER, WALTER TAZEWELL: 1st SHA. Pvt. B. Rockingham Co, ca. 1835. Enl. 6/14/61, Winchester with Co. I, 1st Va. Cav. Captured on picket at Flint Hill, 2/7/62. Exchanged and illegally enlisted/transferred to 1st SHA, 7/30/63. Transferred officially 7/28/64. Paroled 4/15/65, Lynchburg; age 27, 5'5", dark complexion, dark hair, dark eyes. Carpenter, age 35, Pastures Dist., Augusta Co. 1870 census. Receiving pension Augusta Co., 1908, age 78. D. Beckley, W.Va., 1/29/09.

FLANAGAN, WILLIAM W.: 2nd SHA. Pvt. Enl. 12/14/63, Richmond. Transferred to the 1st Engineer Troops 8/3/64. On rolls through (1/20/65). NFR.

FLEINER, JOHN or HERMAN: 2nd SHA. Pvt. Enl. 12/22/62, Port Royal. Wounded in right leg by minnie ball and at Jackson Hosp., Richmond, 11/15/63. Furloughed for 30 days, 12/9/63. Present through (11/3/64). Paroled 5/2/65, at Bowling Green; residence listed as New Orleans with destination New York.

FLIPPIN, JOHN J.: SHA. Postwar reference.

FORD, ADOLPHUS: 2nd SHA. Pvt. Enl. 7/20/63, Richmond. Missing and supposed captured (11/3/64). AWOL (7/24/64). NFR.

FORD, CHARLES EDWARD: 2nd SHA. 1st Lt. B. 11/23/41 at Fairfax Court House, Va. eldest son of Edward R. and Julia F. Ford. One of the early schools he attended was a school taught by Episcopal clergyman, the Rev. R.T. Brown. Entered VMI 8/4/59. Ranked first in first two years, third and fifth respectively in the remaining two. Selected as a drill master to go with others to Richmond to train new Confederate commands. When the armies first closed on Manassas, Ford was at home and on 7/17/61 he was permitted to fall into the ranks of Captain Richardson's company of Col. Kershaw's 2nd S.C. Infantry Regt. Participated in the 1st Battle of Manassas. Assigned to the SHA when it was formed. Promoted to 2nd Lt. 9/22/62, date of rank 8/9/62. On 11/10/63 Ford was promoted to 1st Lt. of artillery in the Provisional Army of the State of Virginia, to rank as of 2/15/63. On 5/25/64 at Hanover C.H. Ford received a minie-ball through the forehead and fell mortally wounded from his horse. Later re-interred in Hollywood Cemetery.

FRAND, _________: 1st SHA. Postwar reference. Killed at Dover, near Aldie.

FREEMAN, E. H.: 2nd SHA. Pvt. Enl. 6/20/64, White Oak Swamp. Present until listed as AWOL (1/20/65). NFR.

FREEMAN, GEORGE S.: 2nd SHA. Pvt. Enl. 3/22/62, Culpeper. Private horse. Captured at Thompson's Crossroads during Stoneman's Raid, 5/4/63 but released and paroled due to lack of transportation for prisoners. At Charlottesville GH with contusio 3/1 - 3/5/64. Absent on horse detail 7/24/64. Captured at Culpeper 12/24/64. Sent to Pt. Lookout via Harper's Ferry. Released on oath 5/18/65. D. 2/26/16, age 76. See obit in Culpeper Exponent, 3/3/16.

FRENCHON, LUCIEN or ANTONIO: 2nd SHA. Pvt. Enl.? Paroled at Winchester 4/30/65. Oath taken 5/24; age 21, 5'10", dark complexion, black hair, hazel eyes.

FRIEMAN, J. C.: 2nd SHA. Postwar reference.

GARDNER, JOHN W.: 1st SHA. Pvt. Enl. 2/27/?, Richmond. In hospital 3/64. Pres. - 4/64. Transferred to Md. Line.

GARIGAN, MICHAEL: 1st SHA. Pvt. Enl. 9/20/62, Darkesville. Pres. Through 4/64. Transferred to Md. Line.

GARRISON, JAMES V.: 1st SHA. Pvt. Enl. 2/14/62, Talladega, Alabama. Abs. On furlough (11/2/63). Present through 8/64. Paroled at Talladega, 6/9/65.Postwar reference states he was injured.

GIBSON, ED.: 1st SHA. Postwar reference.

GIBSON, FAYETTE: 1st SHA. Cpl. Enl. 7/12/62, Richmond. Wagoner 8/9/63 - 10/10/63. Pres. Through 9/14/64 or transferred to Md. Line. Res. of St. Michael's, Talbot Co., Md. D. 12/28/03, buried Loudon Cemetery, Baltimore.

GIBSON, J.: 2nd SHA. Pvt. Clothing receipt only (11/12/64). NFR.

GIBSON, THEODRICK: 2nd SHA. Pvt. Enl. 4/15/62, Richmond. Suffering from carbuncle 10/11 - 10/27/63 at Chimbo #1. Present on rolls through (1/20/65). POW and paroled by 36th Massachusetts at Farmville 4/65. Age 29, Farmer, 1860 Floyd Co. Census.

GIBSON, THOMAS: 1st SHA. Listed in Amos D. Woods' *Floyd County: A History of Its People and Places* as a member of this unit.

GIBSON, W. J. W.: 1st SHA. Pvt. Enl. 8/1/62, Pittsylvania. Pres. through 8/64.

GLEASON, JAMES: 2nd SHA. Pvt. Enl. 5/15/63, Richmond. Camp itch at Chimbo #4, 12/10/63 - transfer to Petersburg GH 1/22/64. Scabies at Williamsburg Episcopal Church Hospital 2/22 - 2/28/64. Deserted (1/20/65). NFR.

GLENN, N. C.: 1st SHA. Pvt. (See 4th Va. Cav.) Enl. 6/22/61, Richmond. At GH # 9 Richmond 6/64. Captured 10/9/64 at Strasburg. Sent to Pt. Lookout. Exchanged 3/28/65.

GOODMAN, B. WILLIAM: 1st SHA. 6th Sgt. Enl. 9/25/62, Darkesville. Detailed to Battalion HQ, 11/2/63. Pres. - 8/64.

GRAY, JAMES M.: 2nd SHA. Pvt. Enl. 4/15/62, Richmond. Listed as wagoner 11/1/62 - 1/1/63. Present through (1/20/65). NFR. Listed in Amos D. Woods' *Floyd County: A History of Its People and Places* as a member of this unit. Age 20, carpenter, 1860 Floyd Co. Census.

GRAY, WILLIAM H.: 2nd SHA. Pvt. Enl. 3/15/62, Franklin Co. At Charlottesville GH, 5/14 - 5/30/64. At GH #9, Richmond 6/6/64 - 6/7. At Chimbo #3 was noted as present with wound through left hand 6/7/64. Furlough for 40 days 6/3? Present through 12/64. Paroled at Charlottesville. NFR. Possibly resident of Floyd Co. in 1860.

GREEN, ZACHARIAH W.: 1st SHA. Pvt. (See Co. E, 11th Va. Cav.) Enl. 4/27/62, Swift Run Gap, Rockingham Co., Va. Transferred to 1st SHA, 2/64? Pres. - 8/64.

GREENWELL, THOMAS W. HERB: 1st SHA. Pvt. Res. of Great Mills, St. Mary's Co., Md. Enl. ?. Captured at Antietam 9/28/62. Exchanged. KIA near Rector's Cross Roads, 6/15/63. Postwar ref. states killed at Aldie, 1863. See also 2nd Md. Battery?

GRIFFIN, MICHAEL C.: 1st SHA. Pvt. Enl. ? Resident of Madison Co., Mississippi, light complexion, brown hair, grey eyes, 5'9". In guard house awaiting sentence of CM for violation of 9th article of war. Received it per G.O. #68, 6/3/63. At Chimbo. #4, 11/22/63. Deserter and captured at Fredericksburg, 9/5/63. Sent to OCP and took Oath 12/13/63, sent north 12/17.

GROVE, WILLIAM H.: 1st SHA. Pvt. (See 12th Va. Cav.) Enl. 3/8/62, Bridgewater. Transferred from Co. H, 12th Va. Cav., (2/64). Pres. - 8/64. Paroled at Staunton, 5/15/65; age 30, 5'9", fair complexion, dark hair, blue eyes.

GRUBBS, W. A.: 1st SHA. Pvt. Enl. 8/63, Richmond. Sick in hospital Chimbo #1 7-8/64. Captured 10/9/64, Columbia Furnace. Sent to Pt. Lookout. Exchanged 3/29/65. May have also served in Co. C, 1st Maryland Cavalry.

GUILLEMOTTE, CHARLES J.P.: 2nd SHA. 1st Sgt-Orderly Sgt. Enl. 4/22/61, New Orleans. Transferred to the SHA. Absent sick (7/64). Abs. Sick (8/64). On returns for the Army District of the Gulf at Mobile, Alabama 9/3/64. Furloughed from SHA 3/7/65, SO#61/4. Present (1/20/65). NFR.

HALL, ______: 2nd SHA. Pvt. Recorded as on guard report 6/3/63. NFR.

HALL, JESSE W.: 1st SHA. Listed in Amos D. Woods' *Floyd County: A History of Its People and Places* as a member of this unit. Age 17, farm laborer, 1860 Floyd Co. Census.

HALLER, URIAH H.: 1st SHA. Pvt. Enl. 9/8/6?, New Market. Pres - 7/64. NFR.

HAMILTON, JAMES C.: 2nd SHA. Pvt. Enl. ? Wounded in foot and at GH Charlottesville 7/4/64 - 8/17. Present (11/64). Deserter (1/20/65). NFR.

HAMMOND, ________: 2nd SHA. Recalled in Shreve's memoir as killed at Fredericksburg. No official records exist for his service.

HARMAN, MADISON: 1st SHA. Listed in Amos D. Woods' *Floyd County: A History of Its People and Places* as a member of this unit. Woods indicates that he was captured and sent to Fort Delaware at some point in the war. Age 32, farmer, 1860 Floyd Co. Census.

HARRIS, ALEXANDER: 1st SHA. Pvt. Enl. 5/15/6?, Richmond. Captured 10/9/64, Strasburg and sent to Pt. Lookout. Exchanged 3/28/65. NFR.

HARRIS, RICHARD HERBERT: 2nd SHA. Postwar reference.

HARRISON, WILLIAM W.: 1st SHA. Pvt. Enl. ? Captured 10/9/64, Strasburg; sent to Pt. Lookout. Exchanged 1/17/65. Present (1/26/65). NFR.

HART, FRANK: 1st SHA. Pvt. Enl. 7/11/6?, Richmond. Wounded 10/11/63. Postwar reference claims he lost his arm at Cold Harbor. 60 day furlough to Mobile, Alabama, 4/11/64. NFR.

HART, S. JAMES: 2nd SHA. Pvt. Clothing receipt only (11/12/64). NFR.

HAYES, HAYDEN: 1st SHA. Pvt. Enl. ? Fever at Chimbo. #2, 5/24/64 - 7/64. Captured 10/9/64, and sent to Pt. Lookout. Died 5/8/65 before released. Buried in POW graveyard #1706.

HAYNES, VIRGIL S.: 1st SHA. Pvt. Enl. 2/4/64, Charlottesville. Captured 10/9/64, sent to Pt. Lookout. Died of chronic diarrhea 2/14/65.

HEDGEMAN, BENJAMIN GRAYSON: 2nd SHA. Pvt. Enl. Co. A, 9th Va. Cav. , transferred 9/20/64 to SHA. Pres. Through (1/20/65). Paroled 4/28/65, Ashland; farmer. NFR. D. Alexandria 12/2/93.

HENDERSON, HENRY: 1st SHA. Pvt. Enl. 2/2/6?, Richmond. Pres. - (11/63). NFR.

HENDERSON, LARKIN T.: 1st SHA. Pvt. Enl. ? At Robertson Hospital, Richmond, 7/22/64, disabled. Pres. (8/64). Transfer to Md. Line revoked.

HENDERSON, MARSHALL VAN BUREN: 2nd SHA. Pvt. B. Louisa Co., 6/23/40. Paroled 4/26/65, Ashland. Residence: Louisa Co. D. Louisa Co., 4/26/13.

HENRY, MATHIAS WINSTON: 2nd SHA. Captain. B. Kentucky, Nov. 28, 1838. Lived in Russellville, Ky. Graduated U.S.M.A., 1860. U.S. Army, 1860-61. 5'10". Lt., Confederate States Cavalry, 2/15/62 (to rank from

3/61). Acting on J.E.B. Stuart's staff, 3/62. Captain of the 2nd Stuart Horse Artillery, 8/9/62. Promoted Maj., 2/21/63. Commanded an artillery battalion in Hood's Division. Transferred to the western theater in 1864. POW at Salisbury, N.C., 4/12/65, released 7/25. Married Susie R. Burwell, 1874. Held Nevada silver mine interests postwar. D. Brooklyn, N.Y., 11/28/1877. Buried in Old Chapel Cemetery, Clarke Co., Va.

HICKLE, JACOB: 1st SHA. Pvt. B. Rockingham Co, ca. 1845. Farmhand, age 15, 1st Dist., Rockingham Co., 1860 census. Enl. 5/22/61, Camp Davis with Co. I, 1st Va. Cav. Detailed to SHA 11/22/61. Died of typhoid at Moore Hospital, Manassas Junction, 1/10/62.

HICKS, GEORGE WASHINGTON: 2nd SHA. Pvt. B. Spotsylvania Co., Va. Enl. 2/1/64, Gordonsville. Teamster 3 - 10/64. Present (1/20/65). Captured at Aberdeen Ch. 4/3/65. Sent to Pt. Lookout. Released 6/13/65; residence: Fredericksburg, fair complexion, light hair, blue eyes, 5'6". NFR. M. Sarah Gordon 1879. D. Fredericksburg 3/16/20 of cancer. Widow applied for pension while residing in Fredericksburg.

HICKS, JOHN L.: SHA. Postwar reference. AWOL.

HICKS, W. GRANVILLE J.: 1st SHA. Pvt. Enl. 2/17/64, Charlottesville. Previous service? Captured at Kelly's Ford 9/19/63, sent to Camp Chase, age 19. Pres. - 8/64. Recd. Cross of Honor at Fredericksburg 1/04.

HIGGINS, JAMES L.: 1st SHA. Cpl. Enl. ? At Chimbo. #1, 10/6/63 - 1/10/64. Captured Gettysburg or Williamsport, 7/4/63. At U.S.A. GH, Baltimore. Cpl. 7/26/64. Another record shows he was captured again? On 10/9/64. Sorrel also killed, $1,866.66 2/3. Exchanged 2/10/65. Paroled 4/23/65, Farmville. Resident of corner of Reid and Charles St., Baltimore. Member of the Harry Gilmore Camp UCV, Towson, Md.

HIGGS, JOHN W.: 2nd SHA. Pvt. Enl. 6/12/63. Transferred to 1st Engineer Corps (Co. I) per SO #181, 8/3/64.

HILL, EDGAR: 1st SHA. 2nd Lt. Enl. 8/3/6?, Richmond. 3rd Cpl. - Sgt. (8/1/64). Paroled at Greensboro, 1865. D. in Louisville, Ky. 12/06/02.

HITT, CHARLES: 2nd SHA. Pvt. Enl. 3/27/62, Culpeper. Private horse. At GH #18, Richmond, 5/23 - 5/30/62. At GH #9, Richmond, 1/10 - 1/11/65. Present (1/20/65). NFR.

HOBSON, DEANE: 1st SHA. Sgt. B. 1840 in Cumberland Co., Va. son of Major Thomas Michael Deane of Ireland and Elizabeth Wallace of Weldon, N.C. Both parents dead by time Deane matriculated at VMI. VMI, '59. In 1861 was working as clerk. Enl. as pvt in Richmond Howitzers. Transferred to Breathed's battery. Enl. 8/1/6?, Richmond. Sick 6/64 at Chimbo. Cpl. 7/17/64. Pres. - 8/64. Paroled 4/13/65, Lynchburg. After the war worked with B & O RR in Maryland and No. Va. Went west in 1878. NFR.

HOLMES, ISAAC: 2nd SHA. Pvt. May have been an original member of ---------the Newtown Artillery, enl. 4/19/61 and present through 10/61. Enl. 12/11/61, Centreville. Detailed at Battalion HQ (1/20/65). Private Horse. NFR.

HOLMES, JOSEPH: 2nd SHA. Sgt. - Pvt. Enl. 12/11/61, Centreville. 4th Sgt. (7/24/64). Pvt. (1/20/65). NFR.

HOPE, THOMAS W.: 2nd SHA. Pvt. Parole record only - 4/65, Ashland; residence: Louisa Co.

HOPKINS, HENRY "Hal": 1st SHA. 5th Cpl. Enl. 7/18/62. WIA, 11/2/62 at Union. Pres. - 4/64. Transferred to Co. K, 1st Va. Cav., 8/14/64. Transferred to Co. K, 1st Md. Cav, 8/15/64. Res. Baltimore, Md.

HOPKINS, WILLIAM FAYETTE: 1st SHA. Enl. 9/1/6?, Richmond. Pres. - 4/64. Postwar ref. states he was wounded near New Baltimore, Va. Transferred to Md. Line?

HORNE, C. G.: 1st SHA. Postwar reference. Md. Soldier's Home.

HORNE, C. S.: 1st SHA. Postwar reference. Md. Soldier's Home.

HOWARD, DARIUS W.: 1st SHA. Listed in Amos D. Woods' *Floyd County: A History of Its People and Places* as a member of this unit. Age 26, merchant, 1860 Floyd Co. Census.

HOXTON, WILLIAM H.: 1st SHA. 2nd Lt. Wounded 2nd Manassas, 8/28 - 8/30/62. Commissioned 1/9/63, to rank 12/30/62. Court of Inquiry convened per SO# 31/3, 3/2/63. Absent wounded 9/ - 10/63. Took leave on 8/5/64. Buried at Alexandria, Va.

HUDGINS, THOMAS D.: 1st SHA. Pvt. One slip in record. NFR.

HUFF, ROBERT E.: 1st SHA. Listed in Amos D. Woods' *Floyd County: A History of Its People and Places* as a member of this unit. Age 15, farm laborer, 1860 Floyd Co. Census.

HUMPHRIES, J. R.: 1st SHA. Sgt. Enl. ? CM for viol. Of 23rd Art. of War, per GO#66, 5/25/63. Awaiting sentence of CM. NFR.

HUNTER, ANDREW: 1st SHA. Pvt. Enl. 3/9/6?, Richmond. Captured 8/11/63. At Ft. McHenry. Apparently exchanged as he was AWOL on 9/6/64. NFR.

HUNTER, JOB: 2nd SHA. Pvt. Postwar reference. Wounded at Brandy Station.

HUNTER, ROBERT: 2nd SHA. Pvt. Enl. 8/1/63, Culpeper. Pres. Until captured 9/14/63, Culpeper. Sent to OCP. Released on oath taken at Washington, D.C., 12/20/63, sent North; resident of Alexandria?, light complexion, brown hair, blue eyes, 5'4". Returned to duty sometime around 7/64. Wounded at "White Tavern below Richmond" in left hand and sent to Jackson Hosp., Richmond 8/20 - 8/27/64. Present through (1/20/65). Another document listed residence as Lawrence CH, S.C. Postwar Alabama pensioner.

IRVING or IRVINE, JESSE B.: 2nd SHA. Pvt. Enl. 6/1/64, Meadow Bridge. Present through (1/20/65). Surrendered at Appomattox. Lived after the war at 110 Edmonson Ave. Baltimore. In Md. Soldier's Home Book.

IRVING or IRVINE, CARTER P.: 2nd SHA. Pvt. Enl. 2/21/61, New Orleans. Transferred to the SHA. Wounded in two fingers on left hand (metacarpal bone) 6/30/64. Sent to Jackson Hospital 7/28/64. Absent sick (1/20/65). Never returned to duty. NFR.

ISAACS, THOMAS: 1st SHA. Pvt. Enl. ? At Chimbo #4, 2/28/63. Wagoner 7/8 - 10/31/63. Detailed to Bttn. HQ - (11/2/63). NFR.

JACOBS, MICHAEL THOMAS: 1st SHA. Pvt. B. Loudoun Co., ca. 1835. Laborer, age 25, Union, Loudoun Co., 1860. Enl. 4/27/61, Union, with Co. H, 1st Va. Cav. Transferred to 1st SHA, 5-6/64. Sent to Hospital with debility, 7/11/64. Died of brain tumor at Chimbo #5, 8/27/64.

JACOBS, W. H.: 2nd SHA. Pvt. Enl. 1/1/64, Charlottesville. At Jackson Hospital, Richmond 7/8/64. Returned to duty 7/8. Absent sick (7/24/64 - 1/20/65). Dysentery at Chimbo #3, 7/30/64. NFR.

JAMES, GEORGE W.: 2nd SHA. Pvt. Enl. 3/1/64, Charlottesville. Absent sick (7/24/64). Pres. (11/13/64). Discharged as a minor, per SO #240/3, 10/9/64. NFR.

JENKINS, THOMAS "Tom": 1st SHA. Pvt. Enl. 6/10/6?, Richmond. Cattle driver, 8 - 10/63. At hospital with rheumatism, 10/16/64 - furloughed for 60 days, 1/28/65. NFR.

JIMISON, JOHN W.: 1st SHA. (See 12th Va. Cav.). Pvt. Enl. 4/13/62, Harrisonburg. Transferred to 1st SHA (2/64). Deserter 10/11/64. NFR.

JOHNSON, FRANK N.: 1st SHA. B. 1847 in Richmond. Pvt. Enl. 7/1/6?, Richmond. Clerk. Oath at Henrico 9/22/65. NFR. In hardware business in St. Louis in 1899.

JOHNSTON, PHILIP PRESTON: 1st SHA. B. King and Queen Co., Va. 4/4/40. Pvt. In 1st Maryland Infantry, 5/17/61. Pvt., 1st Stuart Horse Artillery, spring 1862. Promoted to corporal, sergeant, and on 10/3/62, lieutenant. WIA, 11/2/63. Captain by 2/27/64. Sorrel KIA 9/20/64. Promoted to major by endorsement of Jubal Early 2/8/65. Graduated from Transylvania University, 1868. Married Sallie Chiles, 1870. Judge and breeder of horses in Lexington, Ky. In postwar. Also a member of Ky. Legislature. D. 2/10/25, Lexington, Ky.

JORDAN, EDWARD V.: 2nd SHA. Pvt. Enl. 6/8/61, Campbell Co. Absent on horse detail (7/24/64). Absent with leave (8/9/64). Pres. (11/3/64). Private horse. Pres. - (1/20/65). NFR.

KANE, CHARLES or JAMES C.: 1st SHA. Pvt. Enl. 6/1/6?, Richmond. Pres. - 4/64. Transferred to Md. Line.

KEITH, ADAM: 2nd SHA. Pvt. Enl. 4/26/64, Gordonsville. Absent sick (7/24/64). Died in hospital near Petersburg, 8/64. File claim 6461, $7.00.

KELEHER, P. H.: 1st SHA. Pvt. Enl. ? Wounded right thigh flesh and at Chimbo. #5, 10/9/64. Pres. 8/64. Paroled 4/27/65, Ashland.

KEY, JOHN F.: 1st SHA. Pvt. Enl. 10/1/6?, Martinsburg. Resident of Carroll Co., Md. Captured at Madison CH 12/28/63. Sent to Pt. Lookout. Exchanged 3/65. Paroled 4/27/65.

KING, EDWARD A.: 1st SHA. Sgt. Enl. 7/1/6?, Richmond. Captured at Hagerstown 7/12/63. Sent to Pt. Lookout. Released 5/14/65.

KING, JAMES A.: 2nd SHA. Pvt. Enl. 6/8/61, Campbell. KIA 5/12/64. NFR.

KRISCHMAN, JAMES R.: 1st SHA. Pvt. Enl. 8/18/6?, Richmond. Assigned to SHA 2/64. In hospitals most of the time thereafter, 10/64 - 2/65. Paroled at Lynchburg, 4/15/65.

LACKING, _______: 2nd SHA. Postwar reference. Killed at Brandy Station, 8/1/63.

LANCASTER, ANDREW W.: 1st SHA. Pvt. Enl. 4/16/62, Richmond. Butcher 7 - 10/63. Pres. - 8/64. NFR.

LANIER, SAMUEL: 2nd SHA. Pvt. Enl. 6/5/62, Richmond. Furloughed per S.O. #137/13, 6/9/63. Present through (1/20/65). NFR.

LATHAM, JAMES S.: 1st SHA. Pvt. B. Shenandoah Co., ca.1844. Enl. Co., I, 7th Va. Cav. At Winchester, 8/10/61. Present 11 - 12/61, using a CS horse. Present 2/28/62 - 4/22/62, had his own horse. Re-enl. 3/2/62, Back Creek. Description: age 18, 6', light complexion, sandy hair, and grey eyes. Occupation, farmer. Pres. 4/22/62 - 6/30/62, had his own horse. Present until captured 1/7/63 at Woodstock, while on picket. Sent to Cumberland, Md., then to Wheeling, W.Va., 1/13/63. Confined at the Atheneum Prison, 1/14/63. Desc: age 19, 6', dark complexion, black hair, blue eyes. Transferred to Camp Chase. Paroled 3/28/63 and sent to City Point to be exchanged. Exchanged 4/1/63. Detailed on extra duty in the QM Dept. At Staunton, 7/6/63 - 7/25/63. Transferred to 1st SHA 4/29/64 by order of Gen. R.E.Lee, SO# 111/7. Paroled at Winchester 5/2/65. NFR.

LATIMER, GEORGE: 1st SHA. Pvt. Enl. 4/27/63, Gordonsville. Captured at Lynchburg 4/14/65. Took Oath at Winchester 6/2/65. Residence: Herford Co., Md. Miller's Hotel.

LAWRENCE, JACOB R.: 1st SHA. Listed in Amos D. Woods' *Floyd County: A History of Its People and Places* as a member of this unit. Age 32, farmer, 1860 Floyd Co. Census.

LAWRENCE, SAM: 2nd SHA. Postwar reference.

LESTER, JAMES T.: 2nd SHA. Pvt. SEE ALSO CO. D, 43rd N.C. Enl. 5/20/61, Murfreesboro, N.C. Transferred to the SHA per SO #222/1, 9/16/64. Listed as deserter 2/65. NFR.

LEVY, ELIAS: 2nd SHA. Pvt. Enl. 7/22/63, Richmond. At Chimbo #1 with gonorrhea, 7/17 - 7/27/63. At Chimbo #2 with old wound 7/20 - 9/23/64. From 9/1/64 at GH Charlottesville and GH #9, Richmond - sick from (7/24/64 - 11/64). AWOL (1/20/65). Paroled Lynchburg, 4/15/65. NFR.

LEWIS, JOHN F.: 2nd SHA. Pvt. Enl. 2/1/64, Charlottesville. Teamster. Captured at Aberdeen Church 4/3/65. Sent to Pt. Lookout and released on oath by 6/14/65. Entered Lee Camp Soldier's Home 9/11/14, age 50. D. 5/17/15. Buried in Hollywood Cemetery, Richmond.

LEWIS, WILLIAM G.: 1st SHA. Pvt. Enl. 5/21/6?, Richmond. Wounded (10/11/63) at hospital 11/63. GSW left thigh. At Jackson Hosp. 3/22/64. Captured 10/9/64 to Pt. Lookout. Resident of Savannah, Ga. Mechanic. Oath taken 5/13/65 and released.

LIKENS, JAMES: 1st SHA. Listed in Amos D. Woods' *Floyd County: A History of Its People and Places* as a member of this unit. Age 32, farmer, 1860 Floyd Co. Census.

LINDSEY, ELIJAH: 1st SHA. Pvt. Enl. 3/1/62, Talladega, Alabama. Absent with leave 4/64. Captured 10/9/64 and sent to Pt. Lookout. Exchanged 3/28/65. NFR.

LINDSEY, JAMES S.: 1st SHA. Pvt. Enl. 3/1/62, Talladega, Alabama. Captured 10/9/64, sent to Pt. Lookout. Released on oath, 6/14/65; residence: Talladega, Alabama, florid complexion, brown hair, blue eyes, 5'4"

LINK, JOHN H.: 2nd SHA. Pvt. Enl. 4/18/62, Richmond. Absent sick at Petersburg GH, 4/29 - 5/17/62. Private horse. Absent again (1/20/65). NFR. Listed in Amos D. Woods' *Floyd County: A History of Its People and Places* as a member of this unit. Age 22, field hand, 1860 Floyd Co. Census.

LITTLE, F.H.: 2nd SHA. Pvt. Enl. 6/8/61, Campbell Co. Absent wounded (7/24/64) at GH Petersburg. Still at GH Petersburg from (11/11/64 - 12/29/64). Pres. (1/20/65). NFR.

LLEWELLEN, JOHN: 1st SHA. Pvt. Enl. 3/1/62, Talladega, Alabama. On furlough 2/10/64. Wagoner 8/9 - 10/25/63. Captured at Waynesboro, 3/2/65 and sent to Ft. Delaware. Released on oath, 6/14/65; res: Talladega, dark complexion and hair, grey eyes, 5'7".

LOUDENSLAGER, THOMAS D.: 1st SHA. Pvt. Enl. 9/9/6?, Richmond. Resident of Bainbridge, Ga. GSW right leg 7/15/63. WIA again at Cold Harbor and right arm amputated. Captured at Garrett's Station 4/3/65 and sent to Newport News where he took the oath 6/15/65; light complexion and hair, grey eyes, 5'7".

LUCAS, JOHN T.: 2nd SHA. Pvt. Enl. 4/14/64, Gordonsville. Present through (1/20/65). NFR.

LUCKETT, GEORGE T.: 1st SHA. Pvt. Enl. 7/10/6?, Richmond. Transferred to Md. Line 4/64.

LUMPKIN, JOHN B.: 2nd SHA. Cpl. - Sgt. Enl. 7/1/62, Richmond. Captured 7/2/63 at Westminster. Listed as at Ft. Delaware; resident of Caroline Co., dark complexion, hair, and eyes, 6'3". Captured again?(11/3/64). NFR.

LUSBY, JOHN: 1st SHA. Pvt. Enl. 3/9/62, Richmond. Transferred to Md. Line 4/64.

LUSLY, JAMES: 1st SHA. Postwar reference. B. Baltimore, Md., 1835. Transferred from 1st Md. Infantry. Transferred to 1st Md. Cavalry Bttn. D. Pikesville, Md., 12/5/08.

LYMAN, WILLIAM R.: Staff SHA. Captain of Co. B (2nd), 31st Va. Inf. Near end of war acted as adjutant for Maj. James Breathed and made captain in SHA. Postwar president of the Lincoln Parish Bank, at Ruston, La. Formerly pres. of the Crescent Insurance Co., New Orleans. Chairman for the fund raising for the orphans of Gen. John Hood.

MACKALL, ROBERT McGILL: 1st SHA. Guidon/Color Bearer. B. 2/5/44. Resident of Georgetown, D.C. Enl. 13th Va. Inf., 1861. Transferred to SHA. Enl. 5/10/62, Richmond. Absent sick in hospital GH #26, Richmond, 2/63 & 11/2/63. At Chimbo. #5, 2/18 - 5/18/63. At CSA GH, Charlottesville with gonorrhea 3/17 - 6/1/64. Supposedly assisted Pelham's body after he was mortally wounded. Transferred to Co. K (2nd), 1st Va. Cav., 8/14/64. Transferred to Co. K, 1st Md. Cav., 8/15/64. Enl. Co. D, 43rd Bttn. Va. Cav. (Mosby's), 10/11/64. Paroled Winchester 4/21/65; 5'6", dark complexion, hair and eyes. Member Confederate Veterans Assoc., Washington, D.C. 1922. Resided after the war in Culpeper, Va. D. 6/21/34, Culpeper, Va. Buried in Fairview Cem., Culpeper.

MAHAFFEY, WILLIAM A.: 1st SHA. Pvt. B. Washington Co., Va.; age 22, 6'2", light complexion, grey eyes, dark hair, farmer. Enl. 12/18/61, 1st Va. Cav. Transferred to SHA, 12/18/61. Discharged 8/1/62 due to restricted urethra & chronic gonorrhea. Res. Of Md. D. before 7/4/92.

MALLORY, JAMES OLIVER: 2nd SHA. Postwar reference. Originally w/Co. E., 10th Alabama Inf.

MANGUM, E. J.: 1st SHA. Pvt. Enl. 3/5/6?, Camp Jeb. Possible resident of Durham, N.C. Sick 4/64 - 8/64 when he took a 40 day furlough. NFR.

MANGUM, W. W. or M.: 1st SHA. Pvt. Enl. 3/5/6?, Camp Jeb. AWOL (4/64). Captured 10/9/64 and sent to Pt. Lookout. Resident of Orange Co., N.C. Farmer. NFR.

MANGUS, GEORGE: 1st SHA. Listed in Amos D. Woods' *Floyd County: A History of Its People and Places* as a member of this unit.

MANN, JAMES: 1st SHA. Pvt. Enl. 7/23/6?, Richmond. Wounded 3/7/63, Fredericksburg. In arrest and sentenced to carry log of wood 5/63. Pres. - 8/64. Paroled 4/65, Lynchburg.

MASON, WILLIAM TAYLOR "Willie": 1st SHA. B. Falmouth, Va., youngest son of Dr. Alexander H. Mason of Stafford Co. VMI, '65. Came to Richmond 1863 and joined SHA. Horse killed at Boonsboro Gap, 7/8/63. Commissioned as 1st Lt. and assigned to staff of E.P. Alexander in 1864 (his brother-in-law). Known as Lt. Willie for his young appearance. Burned Farmville Bridge during retreat 4/7/65. D. 9/1867 of yellow fever. Buried in New Orleans. See E. P. Alexander's *Fighting for the Confederacy* for more on Lt. "Willie."

MATHEWS, HENRY H.: 1st SHA. Quartermaster. Enl. 3/16/62, Culpeper. Transferred to Md. Line 4/64. Resided in Pikesville, Md. after the war.

MAYS, STEPHEN: 1st SHA. Pvt. Enl. 1/25/63, Richmond. Captured 7/6/63, Hanover, Pa. Sent to Ft. McHenry & Ft. Delaware. NFR.

McCABE, GEORGE WILLIAM E.: 1st SHA. Pvt. Enl. 7/9/62, Richmond. Wounded and at GH Charlottesville 3/24 - 5/20/64. Transferred to Co. K (2nd), 1st Va. Cav. 8/14/64. Transferred to Co. K, 1st Md. Cavalry, 8/15/64. WIA twice. Also served in 2nd Md. Bttn. Cav.. Resident of Baltimore. D. by 1903.

McCAFFREY, EDWARD: 2nd SHA. Pvt. Enl. 4/20/61, New Orleans. Transferred to the SHA. Wounded 8/27/62, furloughed 10/14/62. At Chimbo #3 with fracture of right leg 9/16 - 11/27/63. At Chimbo #2, 12/2 - 12/15/63. Listed as carpenter from Culpeper Co. Present through (1/20/65). NFR.

McCLELLAN, A. R.: 2nd SHA. Pvt. Clothing receipt only, 11/29/64. NFR.

McCULLARS, JAMES O.: 1st SHA. Pvt. B. Franklin Co., Tennessee. Enl. 2/1/62, Jacksonville, Alabama. Disabled from chronic rheumatism and discharged 8/1/62; age 32, 6'3", dark hair and complexion, grey eyes, farmer.

McGREGOR, JESSE: 2nd SHA. Pvt. Enl. 11/1/61, Alabama. Present through (1/20/65). Paroled at Talladega, Alabama 5/24/65. NFR.

McGREGOR, WILLIAM MORRELL: 2nd SHA. Captain. B. Alabama. Joined SHA 11/14/61 near Centreville, Va. Elected junior 1st lieutenant. Capt. 2/15/63. Leave from 8/21/63 - 11/63. Promoted 2/18/65. Moved from Alabama to Texas after the war. Died at his home in Cameron, Texas.

McLEARN, SAMUEL: 2nd SHA. Pvt. Enl. 8/1/63, Culpeper. Captured 9/13/63, Culpeper. At Sent to Pt. Lookout from OCP. Exchanged 8/64. At Chimbo #9, 6/11/64? Absent sick (1/20/65). On rolls of Barringer's Brigade. Paroled at Greensboro, N.C. 5/2/65. Listed as either a nurse, steward or cook.

McMANNERS, _______: 1st SHA. Postwar reference.

McMULLEN, CHARLES: 1st SHA. Pvt. Enl. 1/16/6?, Richmond. Transferred to Md. Line, 4/64.

McNEEL, WILLIAM: 1st SHA. Pvt. B. Annapolis, Md. Enl. 6/20/62, Richmond. Discharged 11/17/63; clerk, age 27, 5'6½", light complexion, brown eyes, red hair.

McNELLIS, JAMES: 1st SHA. Enl. 7/12/6?, Richmond. In arrest 5/63. Captured 6/30/63, and sent to Ft. Delaware. Postwar reference states he was killed at Fredericksburg.

MILLER, BENJAMIN FRANKLIN: 1st SHA. Pvt. B. on Whistle Creek, Rockbridge Co., 10/8/43. Res. of 4th Dist., Rockbridge Co., 1860 census. Served in Rockbridge Rangers, Co. D, (1st), 10th Va. Cav., 5 - 11/61. Enl. Co. E, 52nd Va. Inf., 5/12/62. WIA (head) at McDowell, 5/8/62. WIA Gaines Mill, 6/27/62. WIA (ankle) Cedar Run, 8/9/62. Transf. to Co. C, 1st Va. Cav., 2/18/64. Transferred to 1st SHA, 7/15/64. Captured 10/9/64 and sent to Pt. Lookout. Exchanged 3/18/65. NFR. Wagonmaker, Natural Bridge Dist., Rockbridge Co., 1870 census. Wheelwright, Buffalo Forge, 1880. Wagonmaker, Sangersville, 1897. Moved to Fordwick. D. at Staunton 10/23/20. Buried in Stonewall Jackson Cemetery, Lexington, Va. "A brave soldier."

MINTZLER or MENTZER, SAMUEL "Sam": 1st SHA. Enl. 9/8/6?, New Market. Transferred to Md. Line. Postwar reference states he was killed at Winchester 10/63.

MONTENEGRO, HENRY: 2nd SHA. Pvt. Enl. 1/1/63, Richmond. Supposed captured (11/3/64). Postwar reference states he deserted and joined 2nd N.Y. Cav.

MOORE, HENRY L.: 2nd SHA. Pvt. Enl. 8/1/61, Norfolk with the 6th Va. Inf. Transferred to the SHA. Present through (1/20/65). Paroled 5/3/65 at Farmville. Rode a light colored sorrel. Distinguished himself at Five Forks and Ream's Station. D. 5/6/19 at his home near Petersburg, Va.

MOORE, JOHN N.: 1st SHA. Pvt. Enl. ? Capt. 5/11/64, Spotsylvania CH. Sent to Pt. Lookout then to Elmira, NY and exchanged 2/25/65. NFR.

MORGAN, PATRICK: 1st SHA. Pvt. Enl. 12/1/6?, Richmond. Captured 7/14/63 or 7/8. NFR.

MORTON, CLEM R.: 1st SHA. Pvt. Enl. 9/15/6?, Richmond. Pres. - 8/64. Paroled 5/15/65. Resided after the war in Baltimore.

MORTON, NICHOLAS SAMUEL MERRYMAN: 1st SHA. Pvt. Enl. 5/14/6?, OCH. Resident of Baltimore. Captured 10/9/64 and sent to Pt. Lookout. Exchanged 3/28/65.

MOYER, JAMES: 2nd SHA. Pvt. Enl. 6/1/64, Richmond. Died at hospital 7/15/64. NFR.

MURDOCK, JAMES CAMPBELL: 1st SHA. Pvt. B. Md., 1840. Attended U. of Edinburg, Scotland. Enl. Leesburg 6/14/61, Co. K (2nd), 1st Va. Cav. Transferred to 1st SHA, 4/62. Appointed 2nd Lt., C.S. Marine Corps 4/8/63. Served on C.S.S. *Richmond*. Ordered to Wilmington, N.C. 5/7/64. Captured at Ft. Fisher, 1/15/65. Exchanged. Assigned to James River Squadron. Reportedly helped to capture the steamers *St. Mary* and *J.B. Spafford* in Chesapeake Bay, 3/31/65. Paroled at Society Hill, S.C., 5/27/65. Res. of Frederick Co., Md. D. 9/10/89, Baltimore. Buried St. Paul's Cemetery.

MURPHY, JOHN: 1st SHA. Pvt. Enl. ? Captured 6/30/63, Westminster. Sent to Ft. Delaware and Pt. Lookout. Exchanged 3/15/65.

MURPHY, PATRICK: 1st SHA. Pvt. Enl. 7/3/6?, Richmond. Captured 7/1/63. NFR.

MURRAY, STERLING: 1st SHA. Orderly Sgt./1st Sgt. Enl. 6/17/6?. Captured 7/2/63. Sent to Pt. Lookout and Fort Columbus, NYH and exchanged 11/1/64. Transferred to Co. K, 1st Md. Cav. Resided in Leesburg, Va. in postwar. Also in Md. Soldier's Home in 1894. D. 10/24/11, age 78.

MURRAY, WILLIAM H.: 1st SHA. Surgeon. Postwar reference.

MUTH, A.: 1st SHA. Postwar reference. Killed at New Baltimore, Va. 10/63.

MYERS, JOHN H.: 1st SHA. Pvt. Enl. 2/9/6?, Richmond. Pres. - 8/64.

NEAL, FRANK: 1st SHA. Pvt. Enl. 9/25/6?, Darkesville. Pres. - 11/64. Transferred to Md. Line?

NEAL, HARRY or HENRY: 1st SHA. Pvt. Enl. ? Deserter and captured 11/29/63 and sent to OCP. Released on oath 3/2/64. NFR.

NELSON, THOMAS F.: SHA. Postwar reference. Soldier's Home, Atlanta, Ga. 12/95.

NESBIT, JAMES: 1st SHA. Pvt. Enl. 1/17/6?, Richmond. Resident of New Orleans, La. In arrest 6/63. At Castle Thunder with bronchitis, 6/20 - 7/11/63. Detailed to QM Dept. Later determined as unfit for duty.

NICHOLSON, A. S.: 2nd SHA. Pvt. Enl. 6/1/64, hanover. Present through (1/20/65). NFR.

NOLLEY, EDWARD (EDMOND) M.: 1st SHA. Listed in Amos D. Woods' *Floyd County: A History of Its People and Places* as a member of this unit. Age 17, 1860 Floyd Co. Census.

NOLLEY, RUFUS F.: 1st SHA. Listed in Amos D. Woods' *Floyd County: A History of Its People and Places* as a member of this unit. Age 21, field hand, 1860 Floyd Co. Census.

NUCKOLS, REUBEN JULIAN: 2nd SHA. Pvt. Enl. Co. G, 4th Va. Cav. 5/9/61. Postwar reference states he acted as a courier for R. E. Lee 11-12/62. Transferred to SHA. Private horse. Postwar reference states he was KIA at Spotsylvania CH 5/9/64. However, he is shown as absent wounded (1/20/65) and receives a parole at Ashland, 4/28/65. Resident of Hanover Co.

O'BRIEN, EDWARD or EDWIN H.: 1st SHA. Pvt. B. 5/13/43. Enl. 6/5/63, Brandy Station. Pres. - 4/64. Transferred to Co. K (2nd), 1st Va. Cav., 8/14/64. Transferred to Co. K, 1st Maryland Cav., 8/15/64. Also served with Mosby and WIA Lewinsville 3/12/65. D. 10/30/99. Buried at Methodist Cem., Alexandria, Va.

O'BRIEN, OWEN: 2nd SHA. Cpl. - Sgt. Enl. 9/4/62, Richmond. 5th Sgt. (1/20/65). Captured at Aberdeen Church, 4/3/65 and later took oath at Pt. Lookout 5/31/65.

O'GRADY, MICHAEL: 2nd SHA. Pvt. Enl. 1/1/63, Essex Co. Detached to Bttn HQ (11/3/64 - 1/20/65). NFR. Entered Lee Camp Soldier's Home 1/5/88, age 70. Discharged 3/21/90.

OWENS, DAVID: 1st SHA. Pvt. Enl. 12/5/61, Camp Alexander. Sent to hospital 12/18/63. AWOL (4/64). NFR.

PAOLI, ALEXANDER or VINCUYO: 2nd SHA. Pvt. Enl. 3/24/63, Richmond. Present through (1/20/65). NFR.

PARKER, THOMAS: 1st SHA. Postwar reference. Killed in Aldie, 6/19/63.

PARKER, W.: 1st SHA. Postwar reference. Buried at Sharon Cem., Middleburg, Va.

PASSIGALLUPPI, JUAN: 2nd SHA. Pvt. Enl. 6/20/62, Richmond. Private horse. Horse killed in battle 9/13/64, Culpeper. $1,485. NFR.

PEARL, JOSEPH: 2nd SHA. Pvt. Enl. 5/1/63, Richmond. At Chimbo #2 with diarrhea 11/27 - 12/5/64. Present through (1/20/65). NFR.

PELHAM, JOHN: SHA. B. Calhoun Co., Alabama, 9/7/38, son of Atkinson and Martha Mumford McGehee Pelham. Attended U.S.M.A. Left prior to graduation in 1861. Lt. of artillery, P.A.C.S., 3/16/61. Served as lieutenant with the Alburtis' Wise Battery, 1861. Commanded as 1st lieutenant of the Newtown Artillery, 10/61 - the reorganization of the battery as the nucleus for the Stuart Horse Artillery Battery. Captain of Pelham's Stuart Horse Artillery, 3/23/62. Major 8/9/62. Posthumous lieutenant colonel (4/4/63) after his being mortally wounded at Kelly's Ford, 3/17/63. Backdated to 3/2/63. Buried in Jacksonville, Alabama. Note: It is interesting that in one of his letters, Captain George Wilmer Brown annotated that John Pelham's middle initial was C. It might have some relevance in that his maternal grandmother was a Clay. John Pelham also had a brother and an uncle named Charles. Brown also listed several other officers and their middle initials are correct with the exception of James Breathed's.

PERALLI, PETER: 2nd SHA. Pvt. Enl. 7/20/62, Richmond. Captured by Kautz's cavalry 9/17/64. Sent to City Point. Joined US service while at Pt. Lookout, 10/14/65. NFR.

PETTY, WILLIAM F.: 2nd SHA. Pvt. Enl. 4/20/62, Culpeper. Pres. Through (1/20/65). NFR.

PHILLIP, JOHN C.: 2nd SHA. Pvt. Enl. 8/13/62, Richmond. Sentenced to perform log drill for a day 5/11/63 for falling asleep on watch. KIA 8/6/64.

PHILLIPS, JOHN ISAAC: 1st SHA. B. 4/17/40 Dorchester Co., Md. Mortally wounded while engaged at Union, Va., one of the caissons exploded 11/2/62. D. 11/3/62.

PHLEGAR, JOSEPH HENRY: 2nd SHA. B. 8/15/1840, Floyd Co., Va. Age 19, 1860 Floyd Co. Census. Pvt. Son of Isaac & Sarah Catherine Rutherford Pflieger. Enl. 4/15/61, Culpeper. Transferred to the SHA. Was noted as one of four recruits that were meant for the 54th Va. Inf. That were diverted by Pelham 4/30/62. Wounded at Fredericksburg, 12/13/62; "operation circular just below elbow same day because of gun shot wound commuting right wrist." At GH #12, Richmond, age 22, occupation listed as farmer. Sent to converted tobacco warehouse in Lynchburg for recuperation. Disabled and at home (11/3/64). Listed as absent wounded (1/20/65). NFR. Applied for pension 4/88; granted $30 annually. D. 1/19/1923, Salem, Va. Buried in Church Hill Cemetery, Salem, Va. Between first wife Angeline Epperly Phlegar and second Rosa Epperly Phlegar. With first wife children were: William Archer, John Thomas, Lila A. M., Eli Emmett Augustus, Isaac L., Rosa A., Mary Margaret, James Patterson Obenshain, and Alice D. Listed in Amos D. Woods' *Floyd County: A History of Its People and Places* as a member of this unit.

PHLEGAR, LEE: 1st SHA. Listed in Amos D. Woods' *Floyd County: A History of Its People and Places* as a member of this unit.

PHLEGAR, N. C.: 1st SHA. Pvt. Enl. 3/15/62, Camp Alexander. Wounded 12/13/62. Same as above?

POFF, ISAAC: 1st SHA. Listed in Amos D. Woods' *Floyd County: A History of Its People and Places* as a member of this unit. Age 24, farmer, 1860 Floyd Co. Census.

POFF, JOSHUA: 1st SHA. Listed in Amos D. Woods' *Floyd County: A History of Its People and Places* as a member of this unit. Age 29, farmer, 1860 Floyd Co. Census.

POFF, WILLIAM H.: 1st SHA. Listed in Amos D. Woods' *Floyd County: A History of Its People and Places* as a member of this unit. Age 24, farmer, 1860 Floyd Co. Census.

POINDEXTER, WILLIAM M.: 1st SHA. Postwar reference. Member of UCV group in Washington, D.C. Alive in 1894.

PORTER, JOHN E.: 1st SHA. Pvt. B. Va., ca. 1847. Enl. Chesterfield Co., 3/8/63 in Co. G, 1st Va. Cav., as substitute for John C. Goode. Transferred to 1st SHA 4/14/64. Pres. - 8/64. NFR. Res. Of Drawbridge, Dorchestr Co., Md. Farmer, age 63, Leigh Dist., Amelia Co. as of the 1910 Va. census.

POWERS, C. W.: 1st SHA. Pvt. B. Franklin Co., Va. Enl. 4/15/62, Richmond. Discharged 7/12/62, "over 35." Age 35, 6', dark complexion and eyes, black hair, farmer.

PRICE, AUGUSTUS: 2nd SHA. Postwar reference.

PRYME, GEORGE A.: 2nd SHA. Pvt.-Cpl. Enl. 5/10/61, Corinth Mississippi in the 12th Mississippi Infantry. Transferred to the SHA. Resident of Vicksburg. 4th Cpl. (1/20/65). Captured at Aberdeen Church 4/3/65 and sent to Pt. Lookout where he was paroled after oath 6/30/65; light complexion, brown hair, grey eyes, 5'61/2". Transportation furnished for Vicksburg.

PUGH, ISAAC: 1st SHA. Listed in Amos D. Woods' *Floyd County: A History of Its People and Places* as a member of this unit. Possibly a resident of Grayson Co.

PUGH, J.: 1st SHA. Listed in Amos D. Woods' *Floyd County: A History of Its People and Places* as a member of this unit. Possibly a resident of Grayson Co.

PURYEAR, G. W.: 2nd SHA. Pvt. Enl. 6/8/61, Campbell. At Chimbo #9, Richmond 8/18/64. At Chimbo #2, granted 40 day furlough for GSW to left heel. 45 days; age 25, carpenter. Absent wounded (1/20/65).

QUEILHE, CLAUDE: 2nd SHA. Pvt. Enl. 3/22/64, Gordonsville. AWOL (7/64). Present again through (1/20/65). NFR.

RASSI, DOMMINICK: 2nd SHA. Postwar reference.

REED, ANDREW J., Sr.: 1st SHA. Listed in Amos D. Woods' *Floyd County: A History of Its People and Places* as a member of this unit. Age 18, 1860 Floyd Co. Census.

REED, JOHN #1: 1st SHA. Listed in Amos D. Woods' *Floyd County: A History of Its People and Places* as a member of this unit. Age 21, 1860 Floyd Co. Census. Brother of Andrew J. Reed.

REED, JOHN #2: 1st SHA. Listed in Amos D. Woods' *Floyd County: A History of Its People and Places* as a member of this unit. Age 44, farmer, 1860 Floyd Co. Census.

REED, ROBERT: 1st SHA. Listed in Amos D. Woods' *Floyd County: A History of Its People and Places* as a member of this unit. Age 29, farmer, 1860 Floyd Co. Census.

REED, WILLIAM B.: 2nd SHA. Pvt. Enl. 4/15/62, Richmond. Detached service in Bttn. QM Dept. (11/3/64 - 1/20/65). NFR. Listed in Amos D. Woods' *Floyd County: A History of Its People and Places* as a member of this unit. Age 24, field hand, 1860 Floyd Co. Census.

REEVES, JOHN F.: 2nd SHA. Pvt. Enl. ? Blacksmith (11/62 - 1/63). Clothes received while at Pratt Hospital, Lynchburg 3/31/64. NFR. Age 37, blacksmith, 1860 Grayson Co. Census. Also served in Co. A, 26th NC Inf. per PWR. Lived in Johnson Co., TN postwar.

REH, W. F.: 1st SHA. Pvt. Bn. Hesse Darmstadt, Germany. Enl. ? Discharged 1/26/63; age 38, 5'7", dark complexion and hair, grey eyes, physician.

RILEY, THOMAS H.: 1st SHA. Pvt. Enl. 9/15/6?, Orange Springs. Pres. - 2/64. NFR. May have gone to Md. Line. D. at Fredericksburg, Va.

ROACH, R. J.: 1st SHA. Pvt. Enl. 5/30/6?, Richmond. Mechanic. Captured 10/9/64 sent to Pt. Lookout. Released 5/14/65.

ROBERTSON, A. J.: 1st SHA. Pvt. Enl. 3/10/6?, CCH. Pres. - 8/64. NFR.

ROBERTSON, ISAAC E.: 1st SHA. Pvt. Enl. 4/15/62, Richmond. Wagoner. Pres. - 8/64. Listed in Amos D. Woods' *Floyd County: A History of Its People and Places* as a member of this unit. Age 26, field hand, 1860 Floyd Co. Census.

ROBERTSON, JESSE: 1st SHA. Pvt. Enl. 5/1/6?, Gordonsville. Pres. - 8/64.

ROE, DAVID: 1st SHA. Pvt. Enl. 3/17/62, Talladega. Paroled at Newton, N.C. by 10th Michigan Cavalry, 4/19/65. Resided in St. Louis after 1900.

ROMAIN, ALEXANDER: 2nd SHA. 4th Cpl. Enl. 12/15/62, Richmond. 1st Cpl? (11/1/64). Present through (1/20/65). NFR.

ROSE, _____: 2nd SHA. Pvt. Enl. ? Only record shows him on guard report 6/15/63. NFR.

ROSE, JESSE: 1st SHA. Pvt. Enl. 6/15/6?, Richmond. Transferred to 1st Md. Cav., 4/64.

ROSSINI, D.F.: 2nd SHA. Pvt. Enl. ? NFR.

RUDDEN, THOMAS: 1st SHA. Pvt. Enl. 12/12/6?, Richmond. Captured 6/30/63. NFR.

RUSSELL, ELIJAH TARLETON: 1st SHA. Cpl. though postwar reference states he was Sgt. Maj. Enl. 7/11/6?, Richmond. Pres. - 4/64 when transferred to Md. Line. Postwar reference states he was killed in Luray Valley.

RUSSELL, G. T.: 2nd SHA. Pvt. Clothes receipts only on file (10/18 & 11/12/64). NFR.

RUSSELL, ROBERT B.: 2nd SHA. Postwar reference. KIA Brandy Station 8/1/63.

RUSSELL, THOMAS ALFRED: 1st SHA. Sgt. B. 1/28/41. Enl. 7/11/6?, Richmond. Present until transferred to Md. Line, 4/64. Enl. Co. F, 43rd Bttn. Va. Cav. (Mosby's), 9/18/64. Captured 10/29/64 at Dulany's near Upperville. Sent to OCP. Transferred to Ft. Warren 2/6/65. Took oath 6/15/65; 5' 10 3/4", dark complexion, brown hair, blue eyes. Res. Easton, Md. Appears in the 1898 reunion of the 43rd in Marshall. Member

Charles S. Winder Camp #989, UCV, Easton, Md. Lived in Orlean 1895 - 1909. D. 12/29/1919. Buried Orlean Cem., Orlean, Va.

RYAN, JAMES A.: 1st SHA. Pvt. Enl. 1/17/6?, Richmond. Mechanic. Detailed as saddler. Captured 10/9/64. Released 5/14/65. Postwar reference states he lost leg at Shady Grove 5/8/64.

RYAN, WILLIAM G.: 2nd SHA. Pvt. Enl. ? at Talladega, Alabama. Rolls show him as discharged 10/2/62; resident of Alabama, age 15, 5'5", dark complexion and hair. Postwar records show him as wounded at Fredericksburg.

RYANS, SAMUEL: 2nd SHA. Pvt. Enl. 11/1/63, OCH. Absent sick (11/3/64) - (1/20/65). Paroled at Talladega, Alabama, 5/24/65. NFR.

SCARBOROUGH, WILLIAM: 2nd SHA. Cpl.-Pvt. Enl. 2/1/62, Alabama. 2nd Cpl. (7/24/64). Pvt. (1/20/65). Captured 4/18/65, Richmond. NFR.

SEYMOUR, CHARLES L.: 1st SHA. Sgt. Enl. 2/14/62, Talladega. Captured 7/2/63. Exchanged 7/31 from Ft. Delaware. Absent sick at home in Alabama, (11/2/63). Pres. - 8/64. NFR.

SHAFER, SAMUEL JACOB: 1st SHA. Pvt. B. Short Hill, Rockbridge Co., 4/19/34. Served in Rockbridge Rangers, Co. D (1st), 10th Va. Cav., 6/61 - 11/61. Enl. Co. C, 1st Va. Cav., Lexington, Va., 3/10/62. WIA (left cheek - ball remained in his head resting on his palate the rest of his life) Kelly's Ford, 3/17/63. RTD and present until WIA Gettysburg 7/3/63. RTD and present until transferred to 1st SHA, 7/25/64. NFR. Farmer, Buffalo Dist., Rockbridge Co., 1870 census. Moved west. Res. Of Texas 1919. D. Topeka, Kansas, 3/1/27. Buried in McFarland Cem., McFarland, Kansas.

SHANKS, DANIEL: 1st SHA. Captain. Lieutenant and commanding guard house 6/63. Leave 12/19/63 & 2/18/64. Wounded by GSW left leg injuring tibia, (11/2/63). Commissioned 5/9/64, to rank 2/27/64. Paroled 5/8/65. Farmer. D. Buried in St. Mary's Co., Md.

SHANKS, MORRIS: 1st SHA. Pvt. Enl. 7/1/64, Reams Station. Pres. - 8/64. Paroled 5/10/65, Beaver Dam, farmer. In Md. Soldier's Home in 1894. Residence shown as 1130 Argyle Ave.

SHAVER, JOHN: 1st SHA. Pvt. Enl. 6/4/62. Contracted typhoid. Discharged /16/62.

SHAW, J. EMMET: 1st SHA. 1st Lt. Commissioned 9/22/62. NFR.

SHEETS, JOHN HARVEY: 1st SHA. B. Rockingham Co., 5/1/42. Farmhand in 1st Dist., Rockingham Co., 18650 census. Enl. In Co. I, 1st Va. Cav. At Harrisonburg, 5/22/61. Detailed temporarily to 1st SHA, 11/22 - 12/31/61. Returned to served with the 1st Va. Cav. Present until captured at Spotsylvania CH 5/12/64. Sent to Pt. Lookout and then Elmira, N.Y. Exchanged 3/14/65. Admitted to Richmond hospital with bronchitis 3/18/65. Furloughed for 30 days 3/24/65. NFR. Receiving pension, Keezletown, Rockingham Co., 1900. Farmer, Central Dist., Rockingham Co., 1910 census. Member S.B. Gibbons Camp, UCV, Harrisonburg.

SHELOR, JAMES F.: 1st SHA. Listed in Amos D. Woods' *Floyd County: A History of Its People and Places* as a member of this unit. Age 26, farmer, 1860 Floyd Co. Census.

SHELOR, WILLIAM MARION: 1st SHA. Artificer. B. Floyd Co., 8/10/39. Enl. 4/16/62, Richmond. Pres. - 8/64. NFR. D. Floyd Co., 11/16/17. Listed in Amos D. Woods' *Floyd County: A History of Its People and Places* as a member of this unit.

SHELTON, GEORGE W.: 1st SHA. Listed in Amos D. Woods' *Floyd County: A History of Its People and Places* as a member of this unit.

SHELTON, JAMES B.: 2nd SHA. Pvt. Clothes receipt only (11/12/64). NFR.

SHEPHERD, JAMES S. TOURO: 1st SHA. 2nd Lt. B. Jefferson Co., 8/21/38. Farmhand, Charlestown PO, Jefferson Co., 1860 census. Enl. Co. B, 1st Va. Cav., Martinsburg, 5/1/61. Present until detailed as hospital attendant 7 - 8/61. Transferred to 1st SHA, 3/23/62. Commissioned 4/62. Actual promotion took place 5/1/62. Present through 9/22/62. NFR. D. 1878. Buried in Shepherd Cem., Shepherdstown, W.Va.

SHIELDS, CHARLES: 2nd SHA "Friend". Pvt. Enl. 12/6/63, OCH. At GH #9 & GH #9 12/9/63 - 1/4/64. Pres. through (1/20/65). NFR.

SHILLING, H.: 2nd SHA. Pvt. Only clothes receipt for 10/18/64. NFR.

SHILLING, JAMES F.: 2nd SHA. Pvt. & Ambulance Driver. Enl. 4/5/62, Richmond. Ambulance driver from 5-10/63. Pres. Through (1/20/65). NFR. Age 15, hired hand, 1860 Floyd Co. Census.

SHILLINGS, FERDINAND: 1st SHA. Listed in Amos D. Woods' *Floyd County: A History of Its People and Places* as a member of this unit.

SHILLINGS, VALENTINE: 1st SHA. Listed in Amos D. Woods' *Floyd County: A History of Its People and Places* as a member of this unit.

SHIRLEY, F.: 2nd SHA. Sgt. Postwar reference.

SHOTWELL, CASWELL: 2nd SHA. Pvt. Enl. Co. C, 4th Va. Cav., 2/1/63. Transferred 3/11/64, Gordonsville. Pres. until absent on furlough (1/20/65). NFR.

SHREVE, GEORGE WASHINGTON: 2nd SHA. Sgt. Enl. 4/15/62, Yorktown. Present through (1/20/65). Paroled at Winchester 4/26/65; age 20, 5'9", fair complexion, dark hair, blue eyes, residence in Fairfax Co.

SHREVE, RICHARD S. "Dick": 2nd SHA. 2nd Lt.-1st Lt. See record as Sgt. Co. G, 8th Va. Inf. Requested by J.E.B. Stuart for a commission 4/2/64. To rank 11/20/63. Paroled at Winchester 4/26/65; age 25, 5'10", dark complexion, black hair, hazel eyes, residence Fairfax Co. Postwar reference stated that he was killed by lightening.

SHURLEY, C. P.: 2nd SHA. 3rd Sgt. Enl. 6/13/62, Richmond. Pres. through (1/20/65). NFR.

SHURLEY, WILLIAM: 2nd SHA. Pvt. Enl. 9/15/63, Culpeper. Pres. through (1/20/65). NFR.

SIMMONS, ROLEY (RHOLIN) M.: 1st SHA. Listed in Amos D. Woods' *Floyd County: A History of Its People and Places* as a member of this unit. Age 34, farmer, 1860 Floyd Co. Census.

SIMPSON, D. P.: 2nd SHA. Postwar reference.

SIMPSON, NAT V. "Beau": 2nd SHA. Pvt. Enl. 9/24/62, White Post. At GH Charlottesville 5/2 - 5/13/63 with debility. At Jackson Hospital 6/14/63. Discharged per S.O. #247/2, 10/16/64. Non-Va. Resident. Postwar reference shows he transferred to Mosby's Command, Co. H. Records of the 43rd Bttn Va. Cav. agree on his service. Paroled 4/23/65 at Winchester. Age 19, 5'9", fair complexion, light hair, grey eyes. Res. Fauquier Co.

SIMPSON, WILLIAM A.: 2nd SHA. Pvt. Enl. 4/15/62, Richmond. Pres. - (1/20/65). NFR. Age 23, farm laborer, 1860 Floyd Co. Census.

SINCLAIR, JAMES: 2nd SHA. Pvt. Enl. 4th Va. Cav. Transferred to the SHA. Detailed with Captain Oden (1/20/65). Recommended as Asst. Quartermaster by McGregor, 3/17/65. Surrendered and paroled at Fairfax C.H. 5/5/65; age 45, grey hair, hazel eyes, dark complexion.

SISSON, C. R. "Kit": 1st SHA. Pvt. Enl. 6/16/6?, Richmond. Transferred to Md. Line 4/64.

SLACK, T. A.: 1st SHA. Pvt. Enl. 6/1/6?, Richmond. Horse killed near Yellow Tavern 5/11/64; $1,633.33. Transferred to Md. Line?

SLEDD, ETHELBERT: 1st SHA. Listed in Amos D. Woods' *Floyd County: A History of Its People and Places* as a member of this unit.

SLUSHER, LEWIS HAMILTON: 2nd SHA. Pvt. Enl. 4/15/62, Richmond. Private horse. Pres. - (1/20/65). NFR. Listed in Amos D. Woods' *Floyd County: A History of Its People and Places* as a member of this unit. Age 28, farmer, 1860 Floyd Co. Census.

SMITH, H. R.: 1st SHA. Sgt. Enl. 3/14/6?, Richmond. Sgt. (7/13/64). Horse killed 9/19/64, $2.333.33.

SMITH, JAMES W.: 1st SHA. Listed in Amos D. Woods' *Floyd County: A History of Its People and Places* as a member of this unit. Age 35, farmer, 1860 Floyd Co. Census.

SMITH, JOHN C.: 2nd SHA. Bugler. Known as "Fighting Jim." Enl. 9/1/62, Richmond. Private horse. At Chimbo #2 with dysentery 11/27 - 12/5/64. Pres. - (1/20/65). NFR.

SMITH, JOHN H. B.: 2nd SHA. Pvt. Enl. ? Paroled 5/15/65, Staunton; age 19, 5'?", fair complexion, light hair, hazel eyes.

SMITH, JOHN THOMAS: 2nd SHA. SEE ALSO Co. D, 43rd N.C. Pvt. Enl. ? Clothing receipts show first record of him with the company 10/64. Pres. - (1/20/65). NFR.

SMITH, JOHN: 1st SHA. Pvt. Enl. 3/1/6?, Richmond. Captured 7/8/63 and sent to Pt. Lookout. Exchanged 3/17/64. Paroled 5/13/65, Staunton.

SMITH, JOSIAH: 2nd SHA. Pvt. Enl. 4/15/62, Richmond. Pres. - (1/20/65). NFR. Listed in Amos D. Woods' *Floyd County: A History of Its People and Places* as a member of this unit. Age 30, farmer, 1860 Floyd Co. Census.

SMITH, L. J.: 1st SHA. Pvt. Enl. 2/14/6_, Richmond. WIA, 10/8/64. D. 10/16/64.

SMITH, LEONIDAS: 2nd SHA. Pvt. Enl. 2/1/64, Charlottesville. Teamster. Pres. - (1/20/65). Paroled at Beaver Dam Station, 5/5/65; farmer from Louisa Co. NFR.

SMITH, R. W.: 2nd SHA. Pvt. Clothes receipt only (11 & 12/64). NFR.

SMITH, W.: 2nd SHA. Clothes receipt only. (11/12/64). NFR.

SMITH, WALTER or WILLIAM GEORGE: 1st SHA. Pvt. Enl. 5/14/6?, OCH. Captured 10/9/64. Released from Pt. Lookout on oath; res: Baltimore, dark complexion, black hair, blue eyes, 5'6." In Md. Soldier's Home in 1894 (residence listed as 1507 Linden Ave.)

SMITH,: 1st SHA. Postwar reference. Washington, D.C. Killed at Tom's Brook, 10/9/64.

SPICER, JOSEPH: 2nd SHA. Pvt. Enl. 1/1/64, Charlottesville. Private horse. AWOL (1/20/65). Captured at Aberdeen Church 4/3/65. Took oath 6/19/65; resident of Culpeper Co., fair complexion, light hair, blue eyes, 5'11 1/2".

SPOLAROUSSA, J. B. or PASLOQUE: 2nd SHA. Pvt. Enl. 7/20/62, Richmond. At GH Charlottesville with acute rheumatism 12/31 - 2/9/64. Pres. through (1/20/65).

SPROWL, A. B.: 1st SHA. Pvt. Pay record only. NFR.

STANLEY, PATRICK: 1st SHA. Pvt. Enl. 7/14/62, Richmond. Deserted 9/29/62. Pres. - 4/64. NFR.

STEPHENS, R. H.: 2nd SHA. Pvt. Enl. 6/8/61, Campbell Co. Injured by fall from horse 8/20/64 and sent to Chimbo #2. Granted 35 day furlough. Private horse. Absent wounded without leave (1/20/65).

STRANEY, EDWARD: 1st SHA. Pvt. Enl. 8/1/6?, Richmond. Detailed to Richmond by Secretary of War 1/64. When off detail in 11/64 he deserted. Took oath at City Point 11/5/64. Sent to Ft. Monroe. Resident of Harpers Ferry, dark complexion, black hair, blue eyes, 5'10½".

STRATTON, D. W.: 1st SHA. Pvt. Enl. 6/13/6?, Richmond. Pres. - 8/64. Paroled at Manchester, 4/22/65.

STREET, CHARLES: 1st SHA. Pvt. Enl. 5/27/6?, Hanover CH. Pres. - 8/64. NFR.

STRONG, JOHN: 2nd SHA. Pvt. Enl. 2/1/64, Charlottesville. Present through (1/20/65). Captured at Aberdeen Church 4/3/65. Sent to Pt. Lookout. Released on Oath 6/19/65; resident Louisa Co., dark complexion and hair, hazel eyes, 5'8 1/2".

SULLY, EDWARD: 2nd SHA. Pvt. Enl. ? Court Martialed per G.O.#66, 5/25/63. At Chimbo #1, with debility 11/30 - 12/10/63. Alive 1900. Member of R. E. Lee Camp UCV #1. Transferred to Co. E, 1st Engineers per SO #45/9, 2/16/64. NFR.

SUMNER, T.: 1st SHA. Pvt. Enl. 4/1/6?, Gordonsville. Pres. - 8/64. NFR.

SUPINGER, JACOB A.: 1st SHA. Pvt. B. Va., ca. 1824. Enl. In Co. C, 1st Va. Cav. At Lexington, Va., 3/10/62. Detailed as teamster until transferred to 1st SHA, 7 - 8/64. Captured 10/9/64 and sent to Pt. Lookout. Released on oath 5/14/65; residence: Rockbridge Co., carpenter. Age 46, Natural Bridge Dist., Rockbridge Co., 1870 census. D. near Summers 2/2/91.

SWANCOAT, WILLIAM "Tom": 1st SHA. Sgt. Enl. ? Resident of Wilkshire Co., Ms. Left sick in enemy lines and capt. 7/13/63, Funkstown and sent to USA GH, Frederick, then to Ft. McHenry. Exchanged 3/17/64. Cpl. (7/64). Paroled 7/27/65, Gordonsville; destination Baltimore, 5 Pearl St.

TALIAFERRO, JAMES PIPER: SHA. Postwar reference. B. 1847

TALIAFERRO, W. G.: 1st SHA. QM Sgt. Enl. 6/17/6?, Richmond. QM Sgt (7/13/64). Pres. - 8/64. NFR.

TAPP, WILLIAM B.: 2nd SHA. Pvt. Enl. 3/27/62, Culpeper. Private horse. Captured at Thompson's Crossroads by Stoneman's cavalry raid 5/4/63. Released and paroled due to lack of transportation to carry prisoners. At Chimbo #1 with Laryngitis 10/18 - 11/18/64. Captured again at Amelia CH 4/4/65. Sent to Pt. Lookout. Released 6/8/65. NFR.

TERRIBERRY, J. S.: 1st SHA. Pvt. Enl. 3/11/6?, Richmond. Pres. - 8/64. Paroled at Talladega, 3/20/65. NFR.

TERRY, GEORGE R.: 1st SHA. Cpl. Enl. 2/1/62, Talladega. Cpl. 7/17/64. NFR. Alive in 1907.

TERRYBERRY, WILLIAM: 1st SHA. Postwar reference.

THOMAS, J. A.: 1st SHA. Pvt. From Lynchburg. Enl. 1/12/6?, Lynchburg. Pres. - 8/64.

THOMAS, JAMES HARRY: 1st SHA. Pvt. Enl. 1/12/64, Charlottesville. Absent in arrest 8/64. NFR.

THOMAS, J. HENRY: 1st SHA. Sgt. Enl. 5/20/6?, Richmond. Pres. - 8/64.

THOMAS, PAULIUS: 1st SHA. Postwar reference.

THORNTON, FRANK A.: 1st SHA. Pvt. Enl. 5/11/6?, Richmond. Transferred to Md. Line.

TIMBERLAKE, PHILIP: 1st SHA. Pvt. Enl. ? Paroled 5/15/65, Louisa CH.

TONGE, RICHARD H. "Rich": 1st SHA. Cpl. Enl. 6/27/62, Richmond. Transferred to Co. K (2nd), 1st Va. Cav., 8/14/64. Transferred to Co. K, 1st Md. Cav., 8/15/64. Res. Of Baltimore. D. by 1903.

TONGE, WILLIAM G. D.: 1st SHA. Postwar reference. B. 1843

TRIPLETT, HAYWOOD FOOTE: 1st SHA. Pvt. B. 1/18/11, Gainesville, Va. Enl. 3/15/6?, Culpeper. Wounded 10/14/63 from shell at Bull Run, and left in enemy lines. Right leg amputated by Surgeon William Murray. Retired to Invalid Corps, 10/12/64. Paroled 5/65 at Bowling Green with destination as home in Price William Co. See obit. in Fredericksburg Star, 1/23/11, p.2, c.1.

TRIPPLET, GEORGE: 1st SHA. Postwar reference.

TRUST, GEORGE: 1st SHA. Pvt. Enl. 4/1/6?, Culpeper. Captured and under the name Henry Withers. Took oath at Ft. Warren, boston, res: Baltimore, grey eyes, brown hair, dark complexion, 5'5". In Md. Soldier's Home in 1894. Residence listed as 39 E. Hill St.

TURNER, GEORGE: 1st SHA. Postwar reference.

TURNER, THOMAS H.: 2nd SHA. Pvt. B.12/26/41. Enl. 1/1/63, Essex Co. Present through (1/20/65). Captured and paroled 5/27/65, Charlottesville. Alive in 1920 at Culpeper Reunion. D. 12/30/24. Buried in Culpeper City Cemetery, Va.

TURNER, WILSON H. P.: 1st SHA. Sgt. Postwar reference. Killed at Groveton, 8/29/62.

TUTT, L. PHILIP: 2nd SHA. Pvt. Enl. 3/25/64, Charlottesville. On horse detail (7/64). Present through (1/20/65) when he was reported as AWOL. NFR.

VERBAL, C. M.: 1st SHA. Pvt. Enl. 2/1/6?, Madison CH. Pres. - 8/64.

VEST, HARVEY L.: 1st SHA. Listed in Amos D. Woods' *Floyd County: A History of Its People and Places* as a member of this unit.

VEST, HENRY L.: 1st SHA. Listed in Amos D. Woods' *Floyd County: A History of Its People and Places* as a member of this unit. Age 25, farmer, 1860 Floyd Co. Census.

VEST, JACOB: 1st SHA. Listed in Amos D. Woods' *Floyd County: A History of Its People and Places* as a member of this unit. Age 22, farmer, 1860 Floyd Co. Census.

VEST, GEORGE WASHINGTON: 1st SHA. Listed in Amos D. Woods' *Floyd County: A History of Its People and Places* as a member of this unit. Age 20, farmer, 1860 Floyd Co. Census.

VINNE, PETER: 2nd SHA. Pvt. Enl. 6/10/62, Richmond. At Winder Div. #5 hospital 11/16 - 11/27/62. Wounded 7/22/63. Postwar reference stated "blown from muzzle of gun at Funkstown." At Chimbo #1 until transferred to Petersburg GH 10/14/63. At Charlottesville GH 2/20/64 - 3/11/64. At Danville GH with wound 6/16/64 - 6/21. Transferred to 2nd N.C. Hospital, Columbia, S.C. 7/2/64. Listed as disabled in hospital (11/64). Captured 4/3/65, Richmond. Sent to Libby Prison, Richmond. Sent to Pt. Lookout and released by 6/6/65.

WADE, JOHN C.: 2nd SHA. Pvt. Only mentioned as discharged 7/12/62; from Floyd Co., age 35, 5'9", dark complexion and hair, gray eyes, farmer. "Over 35." Listed in Amos D. Woods' *Floyd County: A History of Its People and Places* as a member of this unit. Age 41, farmer, 1860 Floyd Co. Census.

WAGNER, HARRY: 1st SHA. Pvt. Postwar reference. Wounded at Beverly Ford, 6/9/63. Resident of Md. Soldier's Home.

WALTERS, WILLIAM: 1st SHA. Listed in Amos D. Woods' *Floyd County: A History of Its People and Places* as a member of this unit. Age 25, farmer, 1860 Floyd Co. Census.

WARD, FRANKLIN: 1st SHA. Pvt. Postwar reference. Age 14, 1860 Grayson Co. Census.

WARD, WILLIAM W.: 1st SHA. Pvt. Enl. 3/25/6?, Richmond. Transferred to Md. Line? Paroled as a deserter 3/8/65, transportation furnished to Washington, D.C. Age 14, 1860 Grayson Co. Census.

WARRO, JOSEPH: 1st SHA. Cpl. Enl. 7/62, Richmond. Captured 7/2/63, Carlisle, Pa. Exchanged. Captured again 5/8/64, Wilderness. Exchanged 11/1/64.

WATKINS, WILLIAM CURTIS: 1st SHA. Sgt. Enl. 4/16/62, Richmond. From Floyd Co. At GH, Liberty, Va., 10 - 12/62 (nurse). Captured 10/9/64 and exchanged from Pt. Lookout, 2/10/65. Listed in Amos D. Woods' *Floyd County: A History of Its People and Places* as a member of this unit. Age 22, farmer, 1860 Floyd Co. Census.

WATKINS, WILLIAM S.: 1st SHA. Pvt. At Chimbo #3, 8 - 11/63. Listed as POW (2/64). AWOL (4/64).

WATTERS, _____: 2nd SHA. Postwar reference. Killed at Chancellorsville.

WEEKS, HENRY CRABTREE: 1st SHA. Pvt. Enl. 12/18/62, Richmond. Captured 7/12/63, Hagerstown. Sent to Pt. Lookout. Exchanged 3/17/65. Transferred to Co. K (2nd), 1st Va. Cav., 8/14/64. Transferred to Co. K, 1st Md. Cav., 8/15/64. NFR.

WEGNER, H. F.: 1st SHA. Pvt. Enl. 11/21/6?, Richmond. Pres. - 4/64. Transferred to Md. Line?

WELSH, JOSEPH or THOMAS C.: 1st SHA. Pvt. Enl. 4/27/6?, Swift Run Gap (11th Va. Cav., Co. E). In arrest 5/63 - 6/63. Violated 9th Art. of War. NFR.

WESTBROOK, J. E.: 2nd SHA. Pvt. Clothes issued 10 - 12/64. Paroled at Richmond 5/8/65. NFR.

WHARTON, GEORGE: 1st SHA. Pvt. Transferred from 1st Va. Cav. 7/28/64. NFR.

WHARTON, WILLIAM T.: 2nd SHA. Sgt. - Pvt. Enl. ? Absent detached (8/9/64 - 1/20/65). Bttn. Commissary Sgt. Paroled at Winchester 4/17/65; age 34, 5'9", dark complexion, black hair, grey eyes. D. Gainesville, Va. 10/1917, age 86. See obit in Fredericksburg Star 10/15/17, p.2, c.5.

WHEELER, WILLIAM WALKER: 2nd SHA. Pvt. B. Augusta Co., 10/1/48. Only a receipt for clothing in files (11/29/64). D. Charlottesville, 12/27/25 or 12/27/27.

WHITE, PATRICK: 1st SHA. Pvt. Enl. 5/27/6?, Richmond. Wounded in the hip and at Chimbo. #2, 11/8/64. Age 46. From Savannah, Ga.?

WICHES or WICKES, HARRY: 1st SHA. Res. Of Md. Enl. 1st SHA about 1863. WIA at Beverly Ford, 6/9/63. Transferred to Co. K (2nd), 1st Va. Cav., 8/14/64. Transferred to Co. K, 1st Md. Cav., 8/15/64. Resided after the war in Baltimore.

WIGFALL, FRANCIS HALSEY: 1st SHA. 2nd Lt. Officer of guard detail 5/63. NFR beyond 10/63. Buried near Baltimore, Md.

WILBURN, W. W.: 1st SHA. Pvt. Enl. 5/1/6?, Campbell CH. Contusion of back at Chimbo. #5, 8/9/64. NFR.

WILE or WILD, DANIEL LUTHER: 1st SHA. Enl. 1/9/6?, Richmond. Pres. - 4/64. NFR. Maryland Line? Resided after the war at Union Station, Baltimore.

WILLIAMS, GEORGE: 2nd SHA. Pvt. Enl. ? At GH Farmville 9/21/64, duty. Absent sick (11/64). Transferred from the Bedford Cavalry (1/20/65). NFR.

WILLIAMS, JAMES R.: 1st SHA. Listed in Amos D. Woods' *Floyd County: A History of Its People and Places* as a member of this unit. Age 39, farmer, 1860 Floyd Co. Census.

WILLIAMS, JOHN: 1st SHA. Listed in Amos D. Woods' *Floyd County: A History of Its People and Places* as a member of this unit. Age 42, farmer, 1860 Floyd Co. Census.

WILLIAMS, Z. F.: 1st SHA. Orderly Sgt. Enl. 5/14/6?, OCH. Detached service on orders of Stuart, 4/64. Sgt. 7/10/64. NFR. In Md. Soldier's Home in 1894. Residence listed as 2233 N. Calvert St. D. ca. 1903.

WILLIS, ROBERT L.: 2nd SHA. Pvt. Enl. 4/15/62, Richmond. Private horse. Present through (1/20/65). NFR. Listed in Amos D. Woods' *Floyd County: A History of Its People and Places* as a member of this unit.

WILLIS, THOMAS E.: 1st SHA. Bugler. Resident of Va. Enl. 11/25/6?, Richmond. Pres. - 8/64. NFR. Age 29, farmer, 1860 Floyd Co. Census.

WILSON, CHARLES: 1st SHA. Enl. 8/1/6?, Richmond. Wounded 5/9/64. NFR.

WIMMER, BURWELL: 1st SHA. Listed in Amos D. Woods' *Floyd County: A History of Its People and Places* as a member of this unit.

WIMMER, HARVEY: 1st SHA. Listed in Amos D. Woods' *Floyd County: A History of Its People and Places* as a member of this unit.

WIMMER, ISAAC: 1st SHA. Listed in Amos D. Woods' *Floyd County: A History of Its People and Places* as a member of this unit.

WIMMER, JACOB: 1st SHA. Listed in Amos D. Woods' *Floyd County: A History of Its People and Places* as a member of this unit.

WINGFIELD, ALFRED: 2nd SHA. Postwar reference.

WINGFIELD, AUSTIN: 2nd SHA. Pvt. Enl. 7/10/62, Richmond. Present through (1/20/65). NFR.

WINGFIELD, LEWIS M.: 1st SHA. Pvt. Enl. 8/7/6?, Richmond. Discharged. Age 20, 5'2", dark complexion and eyes, black hair, butcher.

WINN, RICHARD: 2nd SHA. Postwar reference.

WINSTON, F. (T.) A.: 1st SHA. Listed in Amos D. Woods' *Floyd County: A History of Its People and Places* as a member of this unit. Age 36, cabinet maker, 1860 Floyd Co. Census.

WISE, SAMUEL H.: 1st SHA. Pvt. Enl. ? Paroled 5/16/65; age 27, dark complexion, light hair, blue eyes.

WISE, WILLIAM H.: 1st SHA. Pvt. Enl. ? Paroled 5/16/65; age 24, fair complexion, light hair, grey eyes.

WISEMAN, J. R.: 1st SHA. Pvt. Enl. ? Paroled 5/18/65; age 21, 5'8", light complexion and hair, blue eyes.

WITHERS, HENRY: 1st SHA. (SEE George TRUST). 1st SHA. Pvt. Captured 9/29/64, Washington, D.C. Released from Ft. Warren, Boston, 6/13/64, on oath.

WOOD, THOMAS: 1st SHA. Pvt. Residence: Albemarle Co. Paroled 5/16/65, Charlottesville.

YALLAPPI, JOSEPH FERARA: 2nd SHA. Postwar reference.

YATES, THOMAS FRANK: 1st SHA. Pvt. B. Leonardstown, St. Mary's Co., Md., ca. 1843. Enl. 8/10/63, Richmond. WIA (nose), Carlisle, Pa. 7/1/63. Transferred to Co. K (2nd), 1st Va. Cav., 8/14/64. Transferred to Co. K, 1st Md. Cav., 8/15/64. NFR. Deputy Sheriff, Charles Co., Md. 1871. Printer and Publisher, *St. Mary's Beacon*. Member, Army & Navy Society of Md. Line Association. Entered Old Soldier's Home, Pikesville, Md., 1894 from Leonardstown, Md. Alive in 1903. D. at age 70. Buried in Our Lady's Chapel Cem., St. Mary's Co., Md.

YOUNG, M. B.: 1st SHA. Pvt. Enl. 2/15/63, Richmond. GSW ankle and great toe on right foot, 10/11/63. Furloughed for 40 days from Robertson Hospital, Richmond to Augusta, Ga. Transferred to Cobb's Ga) Legion (2/64) per SO #107/19.

YOUNG, VALENTINE: 2nd SHA. Pvt. Enl. ? At Childersburg, Alabama? At Charlottesville GH with debility 4/22 - 4/27/62. Discharged due to deformity of hand; age 20, 5'10", dark complexion, gray eyes, light hair. From Alabama.

ZIMMERMAN, GEORGE H.: 1st SHA. Pvt. From Maryland. Enl. 12/1/6?, Richmond. An ordained minister. Requested as chaplain for 12th Va. Cav. And 1st Md. Battalion. Transferred.

ZIMMERMAN, WILLIAM E.: 1st SHA. Pvt. Enl. 8/1/6?, Fredericksburg. Transferred to 2nd Bttn. Md. Cavalry 4/64. Resided in Baltimore Co., Md. after the war.

BIBLIOGRAPHY

Manuscripts

Duke University
John Warwick Daniel Papers
Jefferson County Museum, Charles Town, WV.
Roger Preston Chew Collection
Library of Congress
Wigfall Papers
National Archives
Compiled Service Records of the 1st (Shanks') and 2nd (Brown's) Stuart Horse Artillery.
U.S. Army Military History Institute
Robert L. Brake Collection
University of Kentucky
Philip Preston Johnston Papers
University of Virginia
John Warwick Daniel Papers
Virginia Military Institute
Alumni Records
Virginia State Library
Joseph V. Bidgood Papers, Unit Record Department, Adjutant General's Office, Confederate Military Records.

Published Primary Sources

Alexander, Edward Porter. *Fighting or the Confederacy.* Gary W. Gallagher, ed. Chapel Hill, N.C., 1989.

Blackford, W. W. *War Years with Jeb Stuart*. New York, 1945.

Borcke, Heros von. *Memoirs of the Confederate War for Independence*. 2 vols. New York, 1938.

Buck, Samuel D. *With the Old Confeds: Actual Experiences of a Captain in the Line*. Baltimore, 1925.

Cooke, John Esten. *Surry of Eagle's Nest*. New York, 1866.

Davis, George B. *The Official Military Atlas of the Civil War.* Washington, D.C., 1891-1895.

Douglas, Henry Kyd. *I Rode with Stonewall*. Chapel Hill, N.C., 1940.

Gilmor, Harry. *Four Years in the Saddle*. New York, 1866.

Haskell, John Cheves. *The Haskell memoirs: The Personal Narrative of a Confederate Officer by John Haskell*. Gilbert E. Govan & James W. Livingood, editors. New York, 1960.

Hotchkiss, Jedediah. *Make Me a Map of the Valley*. Dallas, 1973.

McCabe, James Dabney. *The Life of Thomas J. Jackson*. Richmond, 1864.

McClellan, H.B. *I Rode with Jeb Stuart*. Bloomington, Indiana, 1959.

McDonald, William N. *History of the Laurel Brigade*. Baltimore, 1907.

Neese, George M. *Three Years in the Confederate Horse Artillery*. Dayton, Ohio, 1988.

Shoemaker, John J. *Shoemaker's Battery: Stuart Horse Artillery, Pelham's Battalion*. Memphis, TN., 1908.

United States War Department. *War of the Rebellion: A Compilation of the Official Records of the Union and Confederate Armies*. 128 vols. Washington, D.C., 1880-1901.

Newspapers and Periodicals

Richmond *Examiner*

Blue and Gray

Civil War Regiments

Civil War Times Illustrated

Confederate Veteran, 1893-1932, 40 vols.

Daily Richmond Enquirer

Exponent, Culpeper, Va.

Free Lance, Fredericksburg, Va.

Jacksonville Republican (Alabama)

Journal of Southern History

London Times

Richmond Sentinel

Richmond Whig

Star, Fredericksburg, Va.

The State, Columbia, S.C.

Southern Historical Society Papers, 1876-1953, 52 vols.

Spectator, Staunton, Va.

Times-Dispatch, Richmond, Va.

Secondary Sources

Andrus, Michael J. *The Brooke, Fauquier, Loudoun, and Alexandria Artillery*. Lynchburg, Va., 1990.

Armstrong, Richard L, *7th Virginia Cavalry*. Lynchburg, Va., 1992.

Beck, Brandon H. and Charles Grunder. *The Second Battle of Winchester*. Lynchburg, Va., 1989.

Brice, Marshall M. *Conquest of a Valley*. Charlottesville, 1965.

Bushong, Millard and Dean M. Bushong. *Fightin' Tom Rosser*. Shippensburg, Pa., 1983.

Carmichael, Peter S. *The Purcell, Crenshaw and Letcher Artillery*. Lynchburg, 1990.

________________. *Lee's Yong Artillerist: William R.J. Pegram*. Charlottesville, Va., 1995.

Carter, III, Samuel. *The Last Cavaliers*. New York, 1979.

Davis, Burke. *Jeb Stuart*. New York, 1957.

Driver, Robert J., Jr. *The 1st and 2nd Rockbridge Artillery*. Lynchburg, 1987.

______________. *The Staunton Artillery - McClanahan's Battery*. Lynchburg, Va., 1988.

______________. *14th Virginia Cavalry*. Lynchburg, Va., 1988.

______________. *1st Virginia Cavalry*. Lynchburg, Va., 1991.

______________. *10th Virginia Cavalry*. Lynchburg, Va., 1992.

______________. *5th Virginia Cavalry*. Lynchburg, 1997.

Evans, Clement Anselm, ed. *Confederate Military History*. 12 vols. Atlanta, 1899.

Freeman, Douglas Southall. *Lee's Lieutenants*. 3 vols. New York, 1942-45.

Hardesty's Historical and Geographical Encyclopedia... Special Virginia Edition. New York, 1884.

Hartzler, Daniel. *Marylanders in the Confederacy*. Westminster, MD 1986.

Hassler, William W. *Colonel John Pelham: Lee's Boy Artillerist*. Richmond, 1960.

Hattaway, Herman. *General Stephen D. Lee*. Jackson, Mississippi, 1976.

Henderson, William D. *The Road to Bristoe Station*. Lynchburg, Va., 1987.

Hennessy, John. *Second Manassas Battlefield Map Study.* Lynchburg, Va.

Hennessy, John. *Return to Bull Run*. New York, 1993.

Keen, Hugh C., and Horace Mewborn. *43rd Battalion Virginia Cavalry: Mosby's Command*. Lynchburg, Va., 1993.

Krick, Robert K. *Lee's Colonels: A Biographical Register of the Field Officers of the Army of Northern Virginia*. Dayton, Ohio, 1991.

Longacre, Edward G. *The Cavalry at Gettysburg*. Lincoln, Nebraska, 1986.

Mahr, Theodore C. *Early's Valley Campaign: The Battle of Cedar Creek*. Lynchburg, Va., 1992.

Matter, William D. *If It Takes All Summer: The Battle of Spotsylvania*. Chapel Hill, N.C., 1988.

Mercer, Philip. *The Gallant Pelham*. Macon, Ga., 1929.

Mickle, William E. *Well Known Confederate Veterans*. New Orleans., 1915.

Milham, Charles G. *The Gallant Pelham: American Extraordinary*. Washington, D.C., 1959

Moore, Robert H., II. *The Charlottesville, Lee Lynchburg, and Johnson's Bedford Artillery*. Lynchburg, 1991.

______________. *The Richmond Fayette, Hampden, Thomas, and Blount's Lynchburg Artillery.* Lynchburg, Va., 1991.

______________. *Chew's Ashby, Shoemaker's Lynchburg, and the Newtown Artillery.* Lynchburg, Va., 1995.

_______________. *Graham's Petersburg, Jackson's Kanawha, and Lurty's Roanoke Horse Artillery.* Lynchburg, Va., 1996.
Musick, Michael P. *6th Virginia Cavalry*. Lynchburg, 1990.
O'Reilly, Frank A. *Stonewall Jackson at Fredericksburg: The Battle of Prospect Hill, December 1, 1862.* Lynchburg, Va., 1993.
Ripley, Warren. *Artillery and Ammunition of the Civil War*. New York, 1970.
Schaff, Morris. *The Spirit of Old West Point*. New York, 1907.
Sears, Stephen W. *Landscape Turned Red*. New York, 1983.
Thomas, Emory M. *Bold Dragoon: The Life of J.E.B. Stuart.* New York, 1986.
Virginia Military Institute. *Virginia Military Institute Register of Former Cadets*. Lexington., 1957.
Walker, Charles D. *Biographical Sketches of the Graduates and Eleves of the Virginia Military Institute*. Philadelphia, 1875.
Wallace, Lee A., Jr. *A Guide to Virginia Military Organizations*. Lynchburg, 1986.
___________. *17th Virginia Infantry*. Lynchburg, Va., 1990.
Weaver, Jeffrey C. *The 1860 Federal Census for Floyd County, Virginia*. Clintwood, Virginia, 1993.
Wert, Jeffry D. *From Winchester to Cedar Creek: The Shenandoah Campaign of 1864*. New York, 1989.
Wise, Jennings C. *Long Arm of Lee*. 2 vols. Lynchburg, 1915.
Wood, Amos D. *Floyd County: A History of its People and Place*. Blacksburg, 1981.
Vandiver, Frank E. *Mighty Stonewall*. New York, 1957.